Math into LaTeX

An Introduction to LaTeX and AMS-LaTeX

*This book is dedicated to those who worked so hard
and for so long to bring these important tools to us:*

The LaTeX3 team
and in particular
Frank Mittelbach (project leader) and David Carlisle

The $\mathcal{AMS}$ team
and in particular
Michael J. Downes (project leader) and David M. Jones

George Grätzer

Math into LaTeX
An Introduction to LaTeX and AMS-LaTeX

BIRKHÄUSER
BOSTON • BASEL • BERLIN

George Grätzer
Department of Mathematics
University of Manitoba
Winnipeg, MB Canada

Library of Congress Cataloging-In-Publication Data

Grätzer, George A.
  Math into LaTeX : an introduction to LaTeX and AMS-LaTeX /
George Grätzer
    p.   cm.
  Includes index.
  ISBN 0-8176-3805-9 (acid-free paper) (pbk. : alk. paper)
    1. AMS-LaTeX.   2. Mathematics printing–Computer programs.
  3. Computerized typesetting.   I. Title.
  Z253.4A65G69 1995   95-36881
  688.2'2544536–dc20   CIP

Printed on acid-free paper   *Birkhäuser*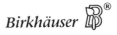
© 1996 Birkhäuser Boston

Design and layout by Merry Sawdey, Minneapolis, MN
Typeset by the author in LaTeX
ISBN 0-8176-3805-9
ISBN 3-7643-3805-9

Printed and bound by Quinn-Woodbine, Woodbine, NJ
Printed in the United States of America

9 8 7 6 5 4 3 2

Short contents

Contents

II Text and math 59

List of tables

List of figures

Preface

It is indeed a lucky author who is given the opportunity to completely rewrite a book barely a year after its publication. Writing about software affords such opportunities (especially if the original edition sold out), since the author is shooting at a moving target.

LaTeX and $\mathcal{AMS}$-LaTeX improved dramatically with the release of the new standard LaTeX (called LaTeX 2_ε) in June of 1994 and the revision of $\mathcal{AMS}$-LaTeX (version 1.2) in February of 1995. The change in $\mathcal{AMS}$-LaTeX is profound. LaTeX 2_ε made it possible for $\mathcal{AMS}$-LaTeX to join the LaTeX world. One of the main points of the present book is to make this clear. This book introduces LaTeX as a tool for mathematical typesetting, and treats $\mathcal{AMS}$-LaTeX as a set of enhancements to the standard LaTeX, to be used in conjunction with hundreds of other LaTeX 2_ε enhancements.

I am not a TeX expert. Learning the mysteries of the system has given me great respect for those who crafted it: Donald Knuth, Leslie Lamport, Michael Spivak, and others did the original work; David Carlisle, Michael J. Downes, David M. Jones, Frank Mittelbach, Rainer Schöpf, and many others built on the work of these pioneers to create the new LaTeX and $\mathcal{AMS}$-LaTeX.

Many of these experts and a multitude of others helped me while I was writing this book. I would like to express my deepest appreciation and heartfelt thanks to all who gave their time so generously. Their story is told in the Afterword.

Of course, the responsibility is mine for all the mistakes remaining in the book. Please send corrections—and suggestions for improvements—to me at the following address:

Department of Mathematics
University of Manitoba
Winnipeg MB, R3T 2N2
Canada
e-mail: George_Gratzer@umanitoba.ca

Introduction

Is this book for you?

This book is for the mathematician, engineer, scientist, or technical typist who wants to write and typeset articles containing mathematical formulas but does not want to spend much time learning how to do it.

I assume you are set up to use LaTeX, and you know how to use an editor to type a document, such as:

```
\documentclass{article}
\begin{document}
The square root of two: $\sqrt{2}$.  I can type math!
\end{document}
```

I also assume you know how to typeset a document, such as this example, with LaTeX to get the printed version:

The square root of two: $\sqrt{2}$. I can type math!

and you can view and print the typeset document.

And what do I promise to deliver? I hope to provide you with a solid foundation in LaTeX, the $\mathcal{AMS}$ enhancements, and some standard LaTeX enhancements, so typing a mathematical document will become second nature to you.

How to read this book?

Part I gives a short course in LaTeX. Read it, work through the examples, and you are ready to type your first paper. Later, at your leisure, read the other parts to become more proficient.

The rest of this section introduces TEX, LaTEX, and $\mathcal{AMS}$-LaTEX, and then outlines what is in this book. If you already know that you want to use LaTEX to typeset math, you may choose to skip it.

TEX, LaTEX, and $\mathcal{AMS}$-LaTEX

TEX is a typesetting language created by Donald E. Knuth; it has extensive capabilities to typeset math. LaTEX is an extension of TEX designed by Leslie Lamport; its major features include

- a strong focus on document structure and the logical markup of text;
- automatic numbering and cross-referencing.

$\mathcal{AMS}$-LaTEX distills the decades-long experience of the American Mathematical Society ($\mathcal{AMS}$) in publishing mathematical journals and books; it *adds to* LaTEX a host of features related to mathematical typesetting, especially the typesetting of multiline formulas and the production of finely-tuned printed output.

Articles written in LaTEX (and $\mathcal{AMS}$-LaTEX) are accepted for publication by an increasing number of journals, including all the journals of the $\mathcal{AMS}$.

Look at the typeset sample articles: `sampart.tex` (in Appendix C, on pages 361–363) and `intrart.tex` (on pages 39–40). You can begin creating such high-quality typeset articles after completing Part I.

What is document markup?

Most word processing programs are WYSIWYG (what you see is what you get); as you work, the text on the computer monitor is shown, more or less, as it'll look when printed. Different fonts, font sizes, italics, and bold face are all shown.

A different approach is taken by a *markup language*. It works with a text editor, an editing program that shows the text, the *source file*, on the computer monitor with only one font, in one size and shape. To indicate that you wish to change the font in the printed copy in some way, you must "mark up" the source file. For instance, to typeset the phrase "Small Caps" in small caps, you type

```
\textsc{Small Caps}
```

The `\textsc` command is a markup command, and the printed output is

SMALL CAPS

TEX is a markup language; LaTEX is another markup language, an extension of TEX. Actually, it's quite easy to learn how to mark up text. For another example, look at the abstract of the `sampart.tex` sample article (page 364), and the instruction

```
\emph{complete-simple distributive lattices}
```

to emphasize the phrase "complete-simple distributive lattices", which when typeset looks like

complete-simple distributive lattices

On pages 364–371 we show the source file and the typeset version of the `sampart.tex` sample article together. The markup in the source file may appear somewhat bewildering at first, especially if you have previously worked on a WYSI-WYG word processor. The typeset article is a rather pleasing-to-the-eye polished version of that same marked up material.[1]

T_EX

TEX has excellent typesetting capabilities. It deals with mathematical formulas as well as text. To get $\sqrt{a^2 + b^2}$ in a formula, type `\sqrt{a^{2} + b^{2}}`. There is no need to worry about how to construct the square root symbol that covers $a^2 + b^2$.

A tremendous appeal of the TEX language is that a source file is *plain text*, sometimes called an ASCII file.[2] Therefore articles containing even the most complicated mathematical expressions can be readily transmitted *electronically*—to colleagues, coauthors, journals, editors, and publishers.

TEX is *platform independent*. You may type the source file on a Macintosh, and your coauthor may make improvements to the same file on an IBM compatible personal computer; the journal publishing the article may use a DEC minicomputer. The form of TEX, a richer version, used to typeset documents is called *Plain TEX*. I'll not try to distinguish between the two.

TEX, however, is a programming language, meant to be used by programmers.

L^AT_EX

LATEX is much easier and safer to work with than TEX; it has a number of built-in safety features and a large set of error messages.

LATEX, building on TEX, provides the following additional features:

- An article is divided into *logical units* such as an abstract, sections, theorems, a bibliography, and so on. The logical units are typed separately. After all the

---

[1] Of course, markup languages have always dominated typographic work of high quality. On the Internet, the most trendy communications on the World Wide Web are written in a markup language called HTML (HyperText Markup Language).

[2] ASCII stands for American Standard Code for Information Interchange.

units have been typed, LaTeX organizes the *placement* and *formatting* of these elements.

Notice line 4 of the source file of the `sampart.tex` sample article

```
\documentclass{amsart}
```

on page 364. Here the general design is specified by the amsart "document class", which is the $\mathcal{AMS}$ article document class. When submitting your article to a journal that is equipped to handle LaTeX articles (and the number of such journals is increasing rapidly), only the *name of the document class* is replaced by the editor to make the article conform to the design of the journal.

- LaTeX relieves you of tedious *bookkeeping chores.* Consider a completed article, with theorems and equations numbered and properly cross-referenced. Upon final reading, some changes must be made—for example, section 4 has to be placed after section 7, and a new theorem has to be inserted somewhere in the middle. Such a minor change used to be a major headache! But with LaTeX, it becomes almost a pleasure to make such changes. LaTeX automatically redoes all the numbering and cross-references.
- Typing the same *bibliographic references* in article after article is a tedious chore. With LaTeX you may use BIBTeX, a program that helps you create and maintain bibliographic databases, so references need not be retyped for each article. BIBTeX will select and format the needed references from the databases.

All the features of LaTeX are made available by the LaTeX format, which you should use to typeset the sample documents in this book.

$\mathcal{AMS}$-LaTeX

The $\mathcal{AMS}$ enhanced the capabilities of LaTeX in three different areas. You decide which of these are important to you.

1. **Math enhancements.** The first area of improvement is a wide variety of tools for typesetting math. $\mathcal{AMS}$-LaTeX provides

- excellent tools to deal with *multiline math formulas* requiring special alignment. For instance, in the following formula, the equals sign $(=)$ is vertically aligned and so are the explanatory comments:

$$
\begin{aligned}
x &= (x + y)(x + z) && \text{(by distributivity)} \\
&= x + yz && \text{(by Condition (M))} \\
&= yz
\end{aligned}
$$

- numerous constructs for typesetting math, exemplified by the following formula:

$$f(x) = \begin{cases} -x^2, & \text{if } x < 0; \\ \alpha + x, & \text{if } 0 \leq x \leq 1; \\ x^2, & \text{otherwise.} \end{cases}$$

- special spacing rules for dozens of formula types, for example

$$a \equiv b \pmod{\Theta}$$

If the above formula is typed inline, it becomes: $a \equiv b \pmod{\Theta}$; the spacing is automatically changed.

- multiline "subscripts" as in

$$\sum_{\substack{i<n \\ j<m}} \alpha_{i,j}^2$$

- user-defined symbols for typesetting math, such as

$$\text{Trunc } f(x), \quad \hat{\hat{A}}, \quad {}^*\!\sum{}^*$$

- formulas numbered in a variety of ways:
  - automatically,
  - manually (by tagging),
  - by groups, with a group number such as (2), and individual numbers such as (2a), (2b), and so on.
- the proof environment and three theorem styles; see the sampart.tex sample article (pages 361–363) for examples.

2. Document classes. $\mathcal{AMS}$-LaTeX provides a number of document classes, including the $\mathcal{AMS}$ article document class, amsart, which allows the input of the title page information (author, address, e-mail, and so on) as separate entities. As a result, a journal can typeset even the title page of an article according to its own specifications without having to retype it.

Many users prefer the visual design of the amsart document class to the simpler design of the classical LaTeX article document class.

3. Fonts. There are hundreds of binary operations, binary relations, negated binary relations, bold symbols, arrows, extensible arrows, and so on, provided by $\mathcal{AMS}$-LaTeX, which also makes available additional math alphabets such as Blackboard bold, Euler Fraktur, Euler Script, and math bold italic. Here are just a few examples:

$$\Leftarrow, \quad \blacktriangle, \quad \sharp, \quad \supsetneq, \quad \mathbb{A}, \quad \mathfrak{p}, \quad \mathcal{E}$$

We have barely scratched the surface of this truly powerful set of enhancements.

What is in the book?

Part I (**Chapter 1**) will help you get started quickly with LaTeX; if you read it carefully, you'll certainly be ready to start typing your first article and tackle LaTeX in more depth.

Part I guides you through:

- marking up text, which is quite easy;
- marking up math, which is not so straightforward (four sections ease you into mathematical typesetting: the first discusses the basic building blocks; the second shows how to build up a complicated formula in simple steps; the third is a formula gallery; and the fourth deals with equations and multiline formulas);
- the anatomy of an article;
- how to set up an article template;
- typing your first article.

Part II introduces the two most basic skills in depth: *typing text* and *typing math*.

Chapters 2 and 3 introduce *text* and *displayed text*. Chapter 2 is very important; when typing your LaTeX document, you spend most of your time typing text. The topics covered include special characters and accents, hyphenation, fonts, and spacing. Chapter 3 covers displayed text including *lists* and *tables*, and for the mathematician, proclamations (theorem-like structures) and proofs.

Chapters 4 and 5 discuss *math* and *displayed math*. Of course, typing math is the heart of any mathematical typesetting system. Chapter 4 discusses this topic in detail, including basic constructs, operators, delimiters, building new symbols, fonts, and grouping of equations. Chapter 5 presents one of the major contributions of $\mathcal{AMS}$-LaTeX: aligned multiline formulas. This chapter also contains other multiline formulas.

Part III discusses the parts of a LaTeX document. In **Chapter 6**, you learn about the structure of a LaTeX document. The most important topics are sectioning and cross-referencing. In **Chapter 7**, the standard LaTeX document classes are presented: `article`, `report`, `book`, and `letter`, along with a description of the standard LaTeX distribution. In **Chapter 8**, the $\mathcal{AMS}$ document classes are discussed. In particular, the title page information for the `amsart` document class and a description of the standard $\mathcal{AMS}$-LaTeX distribution is presented.

Part IV (**Chapter 9**) introduces techniques to *customize* LaTeX to speed up typing source files and typesetting of documents. LaTeX really speeds up with user-defined commands, user-defined environments, and custom formats. You'll learn how parameters that effect the behavior of LaTeX are stored in *counters* and *length commands*, how to change them, and how to design custom lists.

In **Part V** (**Chapters 10 and 11**), we'll discuss two programs: BIBTEX and *MakeIndex* that complement the standard LaTeX distribution; they give a helping hand in making large bibliographies and indices.

Appendices A and B will probably be needed quite often in your work: they contain *math symbol tables* and *text symbol tables*.

Appendix C presents the AMS-LaTeX sample article, sampart.tex, first in typeset form (pages 361–363), then in "mixed" form, showing the source file and the typeset article together (pages 364–371). You can learn a lot about LaTeX and AMS-LaTeX just by reading the source file a paragraph at a time and see how that paragraph looks typeset. Then **Appendix D** rewrites this sample article utilizing the user-defined commands collected in lattice.sty of section 9.5.

Appendix E relates some historical background material on LaTeX: how did it develop and how does it work. **Appendix F** is a brief introduction to the use of PostScript fonts in a LaTeX document. **Appendix G** shows how you can obtain LaTeX and AMS-LaTeX, and how you can keep them up-to-date through the Internet. A work session is reproduced (in part) using "anonymous ftp" (file transfer protocol).

Appendix H will help those who have worked with (Plain) TEX, LaTeX version 2.09, AMS-TEX, or AMS-LaTeX version 1.1, programs from which the new LaTeX and AMS-LaTeX developed. Some tips are given to smooth the transition to the new LaTeX and AMS-LaTeX.

Finally, **Appendix I** points the way for further study. The most important book for extending and customizing LaTeX is *The LaTeX Companion*, the work of Michel Goossens, Frank Mittelbach, and Alexander Samarin [12].

Typographical conventions

To make this book easy to read, I use some very simple conventions on the use of fonts.

Explanatory text is set in the Galliard font, as this text is.

This book is about typesetting math in LaTeX. So often you are told to type in some material and shown how it'll look typeset.

```
I use this font, Computer Modern typewriter style, to show what
you have to type.  All characters have the same width so it's
easy to distinguish it from the other fonts used in this book.
```

I use the same font for commands (`\parbox`), environments (`align`), documents (`sampart.tex`), document classes (`article`), directories and folders (`work`), packages (`amsmath`), counters (`tocdepth`), and so on.

When I show you how something looks when typeset, I use this font, Computer Modern roman, which you'll most likely see when you use LaTeX. This looks sufficiently different from the other two fonts I use so that you should have little difficulty recognizing typeset LaTeX material. If the typeset material is a separate paragraph (or paragraphs), I make it visually stand out even more by adding the little corner symbols on the margin to offset it.

When I give explanations in the text: "Compare iff with iff, typed as `iff` and `if{f}`, respectively." I use the same fonts but since they are not visually set off, it may be a little harder to see that iff is in Computer Modern roman and `iff` is in Computer Modern typewriter style.

Commands are introduced, as a rule, with examples:

```
\\[0.5in]
```

However, sometimes it's necessary to more formally define the syntax of a command. For instance:

```
\\[length]
```

where *length* is a *placeholder*: it represents the length you have to type in. I use the Computer Modern typewriter style italic font for placeholders.

HIN

M

...utive

only two

...t. In this note we prove...t is, complete distri...ngruences.

In this no...

Main Theo...ist
the...trivi...ngru...

2. THE ...

...asic...tion in lattice theo...d un..., see G. A...uhin [2]. ...t with some definitions:

Definition 1. Le...be a complete...tice, and let $p = [u, v]$ be an inter...l of V
Then p is called *complete-prime* if t...llowing three conditions are satisfied:

(M) u is meet-irreduci...but u...t completely...eet-irreducible;

(J) v is join-irreduc...ot completely join-irreducible;

(C) $[u, v]$ is a comp...ice.

Now we prov...

...mma 1. *Let D*...comple...*distributi...lattice satisfying Conditions* (M) *and* (...). *Then $D^{(2)}$ is a sub...lattice o...D^2, hence* $D^{(2)}$ *is a lattice, and $D^{(2)}$ is a comp...te distri...ive lattice satisfying...onditions* (M) *and* (J).

1991 *Mathematics Subject Classification.* Primary: 06B10, Secondary: 06...05.
Key words and phra...Complete...lattice, distributive la...ce, complete...congruence, congruence lattice.
Research supported by the NSF under...nt num...er 23...66.

PART I

A short course

Typing your first article

In this chapter, you'll start writing your first article. All you have to do is to type the (electronic) *source file*; LaTeX does the rest.

In the next few sections, I'll introduce you to the most important commands for typesetting text and math by working through examples. Go to the latter parts of this book for more detail.

The source file is made up of *text*, *math* (for instance, $\sqrt{5}$), and *instructions* to LaTeX. This is how you type the last sentence:

```
The source file is made up of \emph{text}, \emph{math} (for
instance, $\sqrt{5}$), and \emph{instructions} to \LaTeX.
```

In this sentence,

```
The source file is made up of \emph{text}, \emph{math} (for
instance,
```

is text,

```
$\sqrt{5}$
```

is math, and

\emph{ *text* }

is an instruction (a command). Commands, as a rule, start with a backslash \ and are meant to instruct LaTeX; this particular command, \emph, emphasizes *text* given as its *argument* (between the braces). Another kind of instruction is called an *environment*. For instance,

\begin{flushright}

and

\end{flushright}

bracket a `flushright` environment—what is typed inside this environment comes out right justified (lined up against the right margin) in the printed form.

In practice, text, math, and instructions are intertwined. For example,

\emph{My first integral} $\int \zeta^{2}(x) \, dx$

which produces

$$My\ first\ integral \int \zeta^2(x)\, dx$$

is a mixture of all three. Nevertheless, to some extent I try to introduce the three topics: typing text, typing math, and giving instructions to LaTeX (commands and environments) as if they were separate topics.

I introduce the basic features of LaTeX by working with a number of sample documents. If you wish to obtain these documents electronically, create a subdirectory (folder) on your computer, say, `ftp`, and proceed to download all the sample files as described in section G.6. Also create a subdirectory (folder) called `work`. Whenever you want to use one of these documents, copy it from the `ftp` subdirectory (folder) to the `work` subdirectory (folder), so that the original remains unchanged; alternatively, type in the examples as shown in the book. *In this book, the* `ftp` *directory and the* `work` *directory will refer to the directories (folders) you hereby create without further elaboration.*

1.1   *Typing a very short "article"*

First we discuss how to use the keyboard in LaTeX, and then type a very short "article" containing only text.

1.1.1   *The keyboard*

In LaTeX, to type text, use the following keys:

a–z   A–Z   0–9

+   =   *   /   (   )   [   ]

You may also use the punctuation marks

,   ;   .   ?   !   :   '   '   –

and the spacebar, the tab key, and the return (or enter) key.

There are thirteen special keys (on most keyboards):

$ %   &   ~   _   ^   \   {   }   @   " |

used mostly in LaTeX instructions. There are special commands to type most of these special characters (as well as composite characters, such as accented characters) if you need them in text. For instance, $ is typed as \\$, _ is typed as _, and % is typed as \\% (while ä is typed as \\"{a}); however, @ is typed as @. See sections 2.4.4 and 2.4.6 and the tables of Appendix B for more detail.

Every other key is prohibited! (Unless there are special steps are taken; more about this in section 2.1.) Do not use the computer's modifier keys, such as Alt, Ctrl, Command, Option, to produce special characters. LaTeX will either reject or misunderstand them. When trying to typeset a source file that contains a *prohibited character*, LaTeX will display the error message:

```
! Text line contains an invalid character.
l.222 completely irreducible^^?
                            ^^?
```

In this message `l.222` means line 222 of your source file. You must edit this line. The log file (see section 1.11.3) also contains this message.

1.1.2   *Your first note*

We start our discussion on how to type a note in LaTeX with a simple example. Suppose you want to use LaTeX to produce the following:

It is of some concern to me that the terminology used in multi-section math courses is not uniform.

In several sections of the course on matrix theory, the term "hamiltonian-reduced" is used. I, personally, would rather call these "hyper-simple". I invite others to comment on this problem.

Of special concern to me is the terminology in the course by Prof. Rudi Hochschwabauer. Since his field is new, there is no accepted terminology. It is imperative that we arrive at a satisfactory solution.

Create a new file in the work directory with the name note1.tex and type the following (if you prefer not to type it, copy the file from the ftp directory; see page 4):

```
% Sample file: note1.tex
% Typeset with LaTeX format
\documentclass{article}

\begin{document}
It is of some concern to me    that
the terminology used in  multi-section
 math courses is not uniform.

In several sections of the course on
matrix theory, the  term
 ``hamiltonian-reduced'' is used.
  I, personally, would rather call these ``hyper-simple''. I
invite others to comment on this  problem.

Of special concern to me is the terminology in the course
by Prof.~Rudi Hochschwabauer.
  Since his field is new, there is
 no accepted
terminology.   It is imperative
that we arrive at a satisfactory solution.
\end{document}
```

The first two lines start with %; they are *comments* ignored by LaTeX. (The % character is very useful. If, for example, while typing the source file you want to make a comment, but do not want that comment to appear in the typeset version, start the line with %. The whole line will be ignored during typesetting. You can also comment out a part of a line:

```
... % ...
```

The part of a line past the % character will be ignored.)

The line after the two comments names the "document class", which specifies how the document will be formatted.

The text of the note is typed within the "document environment", that is, between the two lines

```
\begin{document}
```

and

```
\end{document}
```

Now typeset `note1.tex`; you should get the same typeset document as shown on page 5.

As seen in the previous example, LaTeX is somewhat different from most word processors. It ignores the way *you* format the text, and follows only the formatting instructions given by the markup commands. LaTeX takes note of whether you put a space in the text, but it ignores *how many spaces* are inserted. In LaTeX, one or more blank lines mark the end of a paragraph. Tabs are treated as spaces. Note that you typed the left double quote as ' ' (two left single quotes) and the right double quote as ' ' (two right single quotes). The left single quote key is not always easy to find; it usually hides in the upper left or upper right corner of the keyboard. The symbol ~ is called a "tie" and keeps `Prof.` and `Rudi` together.

1.1.3 Lines too wide

LaTeX reads the text in the source file one line at a time and when the end of a paragraph is reached, LaTeX typesets it (see section E.2 for a more detailed discussion). Most of the time, there is no need for corrective action. Occasionally, however, LaTeX gets into trouble splitting the paragraph into typeset lines. To illustrate this, modify `note1.tex`: in the second sentence replace "term" by "strange term", and in the fourth sentence delete "Rudi ". Save this modified file with the name `note1b.tex` in the `work` directory. (You'll find `note1b.tex` in the `ftp` directory— see page 4).

Typesetting `note1b.tex`, you get:

It is of some concern to me that the terminology used in multi-section math courses is not uniform.

In several sections of the course on matrix theory, the strange term "hamiltonian-reduced" is used. I, personally, would rather call these "hyper-simple". I invite others to comment on this problem.

Of special concern to me is the terminology in the course by Prof. Hochschwabauer. Since his field is new, there is no accepted terminology. It is imperative that we arrive at a satisfactory solution.

The first line of paragraph two is about 1/4 inch too wide. The first line of paragraph three is even wider. On your monitor, LaTeX displays the message:

```
Overfull \hbox (15.38948pt too wide) in paragraph at lines 10--15
[]\OT1/cmr/m/n/10 In sev-eral sec-tions of the course on ma-trix
the-ory, the strange term ''hamiltonian-
  []
Overfull \hbox (23.27834pt too wide) in paragraph at lines 16--22
[]\OT1/cmr/m/n/10 Of spe-cial con-cern to me is the ter-mi-nol-ogy
```

```
in the course by Prof. Hochschwabauer.
[]
```

You'll find the same message in the log file `note1b.log` (see section 1.11.3).

The reference

```
Overfull \hbox (15.38948pt too wide) in paragraph at lines 10--15
```

is made to paragraph two (lines 10–15); the typeset version has a line (line number unspecified within the typeset paragraph) which is 15.38948pt too wide. LaTeX uses *points* (pt) to measure distances; there are about 72 points to an inch. The next two lines

```
[]\OT1/cmr/m/n/10 In sev-eral sec-tions of the course on ma-trix
the-ory, the strange term ''hamiltonian-
```

identify the source of the problem: LaTeX would not hyphenate

```
hamiltonian-reduced,
```

since it only (automatically) hyphenates a hyphenated word only at the hyphen. You may wonder what `\OT1/cmr/m/n/10` signifies. It says that the current font is the Computer Modern roman font at size 10 points (see section 2.6.1).

The second reference

```
Overfull \hbox (23.27834pt too wide) in paragraph at lines 16--22
```

is made to paragraph three (lines 16–22). The problem is with the word

```
Hochschwabauer
```

which the hyphenation routine of LaTeX can't handle. (If you use a German hyphenation routine, it'll have no difficulty hyphenating Hochschwabauer.)

If you encounter such a problem, try to reword the sentence or add an optional hyphen `\-`, which encourages LaTeX to hyphenate at this point if necessary. For instance, rewrite Hochschwabauer as

```
Hoch\-schwabauer
```

and the second problem goes away.

Sometimes a small horizontal overflow is difficult to spot. The `draft` document class option is very useful in this case: it'll paint an ugly slug on the margin to mark an overfull line; see sections 7.1.2 and 8.4 for document class options. You may invoke this option by changing the `\documentclass` line to

```
\documentclass[draft]{article}
```

You'll find this version of `note1b.tex` under the name `noteslug.tex` in the `ftp` directory.

1.1.4 More text features

Next you'll produce the following note in LaTeX:

September 18, 1995

From the desk of George Grätzer

February 7–21 *please* use my temporary e-mail address:

George_Gratzer@umanitoba.ca

Type in the following source file, save it as note2.tex in the work directory (you'll also find note2.tex in the ftp directory):

```
% Sample file: note2.tex
% Typeset with LaTeX format
\documentclass{article}

\begin{document}
\begin{flushright}
   \today
\end{flushright}
\textbf{From the desk of George Gr\"{a}tzer}\\[10pt]

February~7--21 \emph{please} use my temporary e-mail address:
\begin{center}
   \texttt{George\_Gratzer@umanitoba.ca}
\end{center}
\end{document}
```

This note introduces several additional features of LaTeX:

- The \today command displays today's date.
- Use environments to *right justify* or *center* text. Use the \emph command to *emphasize* text; the text to be emphasized is surrounded by { and }. Use \textbf for **bold** text; the text to be made bold is also surrounded by { and }. Similarly, use \texttt for typewriter style text. \emph, \textbf, and \texttt are examples of *commands with arguments*. Note that command names are case sensitive; do not type \Textbf or \TEXTBF in lieu of \textbf.
- LaTeX commands (almost) always start with \ followed by the command name, for instance, \textbf. The command name is terminated by the first non-alphabetic character.

- Use double hyphens for number ranges (en-dash): 7--21 prints 7–21; use triple hyphens (---) for the "em-dash" punctuation mark—such as the one in this sentence.
- If you want to create additional space between lines (as in the last note under the line **From the desk** ...), use the command \\[10pt] with an appropriate amount of vertical space. (\\ is the newline command—see section 2.7.1; the variant used in the above example is the newline with additional vertical space.) The distance may be given in points, centimeters (cm), or inches (in). (72.27 points make an inch.)
- There are special rules for *accented characters* and some *European characters*. For instance, ä is typed as \"{a}. Accents are explained in section 2.4.6 (see also the tables in Appendix B).

You'll seldom need to know more than this about typing text. For more detail, however, see Chapters 2 and 3. All text symbols are organized into tables in Appendix B.

1.2   Typing math

Now you can start mixing text with math formulas.

1.2.1   The keyboard

In addition to the regular text keys (section 1.1.1), three more keys are needed to type math:

$$| \quad < \quad >$$

(| is the shifted \ key on many keyboards.)

1.2.2   A note with math

You'll begin typesetting math with the following note:

In first year Calculus, we define intervals such as (u, v) and (u, ∞). Such an interval is a *neighborhood* of a if a is in the interval. Students should realize that ∞ is only a symbol, not a number. This is important since we soon introduce concepts such as $\lim_{x \to \infty} f(x)$.

When we introduce the derivative

$$\lim_{x \to a} \frac{f(x) - f(a)}{x - a},$$

we assume that the function is defined and continuous in a neighborhood of a.

To create the source file for this mixed math and text note, create a new document with an editor. Name it `math.tex`, place it in the `work` directory, and type in the following source file—or copy `math.tex` from the `ftp` directory:

```
% Sample file: math.tex
% Typeset with LaTeX format
\documentclass{article}

\begin{document}
In first year Calculus, we    define intervals   such as
$(u, v)$ and $(u, \infty)$.   Such an interval is a
\emph{neighborhood} of   $a$
if  $a$ is in the interval.   Students should
realize that   $\infty$ is only a
symbol, not a number.   This is important since
we soon introduce concepts
 such as $\lim_{x \to \infty} f(x)$.

When we introduce the derivative
\[
   \lim_{x \to a} \frac{f(x) - f(a)}{x - a},
\]
we assume that the function is defined and continuous
in a neighborhood of   $a$.
\end{document}
```

This note introduces the basic techniques of typesetting math with LaTeX:

- There are two kinds of math formulas and environments: *inline* and *displayed*.
- *Inline* math environments open and close with $.
- *Displayed* math environments open with \[and close with \].
- LaTeX ignores the spaces you insert in math environments with two exceptions: spaces that delimit commands (see section 2.3.1) and spaces in the argument of commands that temporarily revert into text mode. (\mbox is such a command; see section 4.5.) Thus spacing in math is important only for the readability of the source file. To summarize:

Rule ■ Spacing in text and math

In text mode, many spaces equal one space, while in math mode, the spaces are ignored.

- The same formula may be typeset differently depending on which math environment it's in. The expression $x \to a$ is typed as a subscript to lim in the inline formula $\lim_{x \to a} f(x)$, typed as `$\lim_{x \to a} f(x)$`, but it's placed below lim in the displayed version:

$$\lim_{x \to a} f(x)$$

typed as

```
\[
    \lim_{x \to a} f(x)
\]
```

- A math symbol is invoked by a command. Examples: the command for ∞ is `\infty` and the command for $\to$ is `\to`. The math symbols are organized into tables in Appendix A.

  To access most of the symbols listed in Appendix A by name, use the amssymb package; in other words, the article should start with

```
\documentclass{article}
\usepackage{amssymb}
```

The amssymb package loads the amsfonts package, which contains the commands for using the AMSFonts (see section 4.14.2).
- Some commands such as `\sqrt` need *arguments* enclosed in { and }. To typeset $\sqrt{5}$, type `$\sqrt{5}$`, where `\sqrt` is the command and 5 is the argument. Some commands need more than one argument. To get

$$\frac{3 + x}{5}$$

type

```
\[
    \frac{3+x}{5}
\]
```

`\frac` is the command, 3 + x and 5 are the arguments.

There are many mistakes you can make, even in such a simple note. You'll now introduce mistakes in math.tex, by inserting and deleting % signs to make the mistakes visible to LaTeX one at a time. Recall that lines starting with % are ignored by LaTeX. Type the following source file, and save it under the name mathb.tex in the work directory (or copy over the file mathb.tex from the ftp directory).

```
% Sample file: mathb.tex
% Typeset with LaTeX format
\documentclass{article}
```

```
\begin{document}
In first year Calculus, we   define intervals  such as
%$(u, v)$ and $(u, \infty)$.  Such an interval is a
 $(u, v)$ and    (u, \infty)$.  Such an interval is a
 {\emph{neighborhood} of $a$
if $a$ is in the interval.  Students should
realize that  $\infty$ is only a
symbol, not a number.  This is important since
we soon introduce concepts
 such as $\lim_{x \to \infty} f(x)$.
%such as $\lim_{x \to \infty f(x)$.

When we introduce the derivative
\[
   \lim_{x \to a} \frac{f(x) - f(a)}{x - a}
   %\lim_{x \to a} \frac{f(x) - f(a)   x - a}
\]
we assume that the function is defined and continuous
in a neighborhood of   $a$.
\end{document}
```

Exercise 1   Note that in line 8, the second $ is missing. When you typeset the mathb.tex file, LaTeX sends the error message:

```
! Missing $ inserted.
<inserted text>
                $
1.8 ..., v)$ and    (u, \infty
                             )$.  Such an interval is a
?
```

Since you omitted $, LaTeX reads (u, \infty) as text; but the \infty command instructs LaTeX to typeset a math symbol, which can only be done in math mode. So LaTeX offers to put a $ in front of \infty. LaTeX suggests a cure, but in this example it comes too late. Math mode should start just prior to (u.

Exercise 2   In the mathb.tex file, delete % at the beginning of line 7 and insert a % at the beginning of line 8 (this eliminates the previous error); delete % at the beginning of line 15 and insert a % at the beginning of line 14 (this introduces a new error: the closing brace of the subscript is missing). Save the changes, and typeset the note. You get the error message:

```
! Missing } inserted.
<inserted text>
                }
```

```
1.15 ...im_{x \to \infty f(x)$
```

```
?
```

LATEX is telling you that a closing brace } is missing, but it's not sure where. LATEX noticed that the subscript started with { and it reached the end of the math formula before finding }. You must look in the formula for a { that is not closed, and close it with }.

Exercise 3   Delete % at the beginning of line 14 and insert a % at the beginning of line 15, which removes the last error, and delete % at the beginning of line 20 and insert a % at the beginning of line 19 (introducing the final error: deleting the closing brace of the first argument of \frac). Save and typeset the file. You get the error message:

```
! LaTeX Error: Bad math environment delimiter.
```

```
1.21 \]
```

There is a bad math environment delimiter in line 21, namely, \]. So the reference to

```
! Bad math environment delimiter.
```

is to the displayed formula. Since the environment delimiter is correct, something must have gone wrong with the displayed formula. This is what happened: LATEX was trying to typeset

```
\lim_{x \to a} \frac{f(x) - f(a)   x - a}
```

but \frac needs two arguments. LATEX found f(x) - f(a)   x - a as the first argument. While looking for the second, it found \], which is obviously an error (it was looking for a {).

1.2.3   *Building blocks of a formula*

A formula is built up from various types of components. We group them as follows:

- Arithmetic
- Subscripts and superscripts
- Accents
- Binomial coefficients
- Congruences
- Delimiters
- Operators
- Ellipses
- Integrals

- Matrices
- Roots
- Sums and products
- Text

Some of the commands in the following examples are defined in the amsmath package; in other words, to typeset these examples with the article document class, the article should start with

```
\documentclass{article}
\usepackage{amssymb,amsmath}
```

Arithmetic The arithmetic operations $a + b$, $a - b$, $-a$, a/b, ab are typed as expected:

```
$a + b$, $a - b$, $-a$, $a / b$, $a b$
```

If you wish to use $\cdot$ or $\times$ for multiplication, as in $a \cdot b$ or $a \times b$, use \cdot or \times, respectively. The expressions $a \cdot b$ and $a \times b$ are typed as follows:

```
$a \cdot b$   $a \times b$
```

Displayed fractions, such as

$$\frac{1 + 2x}{x + y + xy}$$

are typed with \frac:

```
\[
    \frac{1 + 2x}{x + y + xy}
\]
```

The \frac command is seldom used inline.

Subscripts and superscripts Subscripts are typed with _ (underscore) and superscripts with ^ (caret). Remember to enclose the subscripts and superscripts with { and }. To get a_1, type the following characters:

Go into inline math mode:	$
type the letter a:	a
subscript command:	_
bracket the subscripted 1:	{1}
exit inline math mode:	$

that is, type a_{1}. Omitting the braces in this example causes no harm; however, to get a_{10}, you *must* type a_{10}. Indeed, a_10 prints a_10. Further examples: a_{i_1}, a^2, a^{i_1} are typed as

`$a_{i_{1}}$, $a^{2}$, $a^{i_{1}}$`

Accents The four most often used math accents are:

$\bar{a}$ typed as   `$\bar{a}$`

$\hat{a}$ typed as   `$\hat{a}$`

$\tilde{a}$ typed as   `$\tilde{a}$`

$\vec{a}$ typed as   `$\vec{a}$`

Binomial coefficients The amsmath package provides the \binom command for binomial coefficients. For example, $\binom{a}{b+c}$ is typed inline as

`$\binom{a}{b + c}$`

whereas the displayed version

$$\binom{a}{b + c}\binom{\frac{n^2-1}{2}}{n + 1}$$

is typed as

```
\[
    \binom{a}{b + c} \binom{\frac{n^{2} - 1}{2}}{n + 1}
\]
```

Congruences The two most important forms are:

$$a \equiv v \pmod{\theta} \quad \text{typed as} \quad \text{\$a \backslash equiv v \backslash pmod\{\backslash theta\}\$}$$
$$a \equiv v \pod{\theta} \quad\quad \text{typed as} \quad \text{\$a \backslash equiv v \backslash pod\{\backslash theta\}\$}$$

The second form requires the amsmath package.

Delimiters These are parenthesis-like symbols that vertically expand to enclose a formula. For example: $(a + b)^2$, which is typed as `$(a + b)^{2}$`, and

$$\left(\frac{1 + x}{2 + y^2} \right)^2$$

which is typed as

```
\[
    \left( \frac{1 + x}{2 + y^{2}} \right)^{2}
\]
```

contain such delimiters. The \left(and \right) commands tell LATEX to size the parentheses correctly (relative to the size of the symbols inside the parentheses). Two further examples:

$$\left| \frac{a+b}{2} \right|, \quad \|A^2\|$$

would be typed as:

```
\[
  \left| \frac{a + b}{2} \right|,
      \quad \left\| A^{2} \right\|
\]
```

where \quad is a spacing command (see section 4.11 and Appendix A).

Operators To typeset the sine function $\sin x$, type: $\sin x$. Note that $sin x$ prints: $sinx$, where the typeface of sin is wrong, as is the spacing.

LATEX calls \sin an *operator*; there are a number of operators listed in section 4.7.1 and Appendix A. Some are just like \sin; others produce a more complex display:

$$\lim_{x \to 0} f(x) = 0$$

which is typed as

```
\[
    \lim_{x \to 0} f(x) = 0
\]
```

Ellipses The ellipsis (...) in math sometimes needs to be printed as low dots and sometimes as (vertically) centered dots. Print low dots with the \ldots command as in $F(x_1, x_2, \ldots, x_n)$, typed as

```
$F(x_{1}, x_{2}, \ldots , x_{n})$
```

Print centered dots with the \cdots command as in $x_1 + x_2 + \cdots + x_n$, typed as

```
$x_{1} + x_{2} + \cdots + x_{n}$
```

If you use the amsmath package, there is a good chance that the command \dots will print the ellipsis as desired.

Integrals The command for an integral is \int; the lower limit is a subscript and the upper limit is a superscript. Example: $\int_0^\pi \sin x \, dx = 2$ is typed as

```
$\int_{0}^{\pi} \sin x \, dx = 2$
```

\, is a spacing command (see section 4.11 and Appendix A).

Matrices The amsmath package provides you with a `matrix` environment:

$$
\begin{matrix}
a+b+c & uv & x-y & 27 \\
a+b & u+v & z & 134
\end{matrix}
$$

which is typed as follows:

```
\[
   \begin{matrix}
      a + b + c & uv    & x - y & 27\\
      a + b     & u + v & z     & 134
   \end{matrix}
\]
```

The matrix elements are separated by &; the rows are separated by \\. The basic form gives no parentheses; for parentheses, use the `pmatrix` environment; for brackets, the `bmatrix` environment; for vertical lines (determinants, for example), the `vmatrix` environment; for double vertical lines, the `Vmatrix` environment. For example,

$$
\mathbf{A} = \begin{pmatrix} a+b+c & uv \\ a+b & u+v \end{pmatrix} \begin{pmatrix} 30 & 7 \\ 3 & 17 \end{pmatrix}
$$

is typed as follows:

```
\[
   \mathbf{A} =
   \begin{pmatrix}
      a + b + c & uv\\
      a + b & u + v
   \end{pmatrix}
   \begin{pmatrix}
      30 & 7\\
      3 & 17
   \end{pmatrix}
\]
```

Roots \sqrt produces the square root; for instance, $\sqrt{5}$ is typed as

```
$\sqrt{5}$
```

and $\sqrt{a + 2b}$ is typed as

```
$\sqrt{a + 2b}$
```

The *n*th root, $\sqrt[n]{5}$, is done with two arguments:

`$\sqrt[n]{5}$`

Note that the first argument is in brackets [ ]; it's an *optional argument* (see section 2.3).

Sums and products The command for sum is \sum and for product is \prod. The examples

$$\sum_{i=1}^{n} x_i^2 \qquad \prod_{i=1}^{n} x_i^2$$

are typed as

```
\[
    \sum_{i=1}^{n} x_{i}^{2} \qquad \prod_{i=1}^{n} x_{i}^{2}
\]
```

\qquad is a spacing command; it separates the two formulas (see section 4.11 and Appendix A).

Sums and products are examples of *large operators*; all of them are listed in section 4.8 and Appendix A. They display in a different style (and size) when used in an inline formula: $\sum_{i=1}^{n} x_i^2 \quad \prod_{i=1}^{n} x_i^2$.

Text Place text in a formula with an \mbox command. For instance,

$$a = b \quad \text{by assumption}$$

is typed as

```
\[
    a = b \mbox{\qquad by assumption}
\]
```

Note the space command \qquad in the argument of \mbox. You could also have typed

```
\[
    a = b \qquad \mbox{by assumption}
\]
```

because \qquad works in text as well as in math.

If you use the amsmath package, then the \text command is available in lieu of the \mbox command. It works just like the \mbox command except that it automatically changes the size of its argument as required, as in a^{power}, typed as

```
$a^{ \text{power} }$
```

If you do not want to use the large amsmath package, the tiny amstext package also provides the \text command (see section 8.5).

1.2.4 *Building a formula step-by-step*

It is simple to build up complicated formulas from the components described in section 1.2.3. Take the formula

$$\sum_{i=1}^{\left[\frac{n}{2}\right]} \left(x_{i,i+1}^{i^2} \atop \left[\frac{i+3}{3}\right] \right) \frac{\sqrt{\mu(i)^{\frac{3}{2}}(i^2-1)}}{\sqrt[3]{\rho(i)-2}+\sqrt[3]{\rho(i)-1}}$$

for instance. You should build this up in several steps. Create a new file in the work directory. Call it formula.tex and type in the lines:

```
% File: formula.tex
% Typeset with LaTeX format
\documentclass{article}
\usepackage{amssymb,amsmath}
\begin{document}
\end{document}
```

and save it. At present, the file has an empty document environment.[1] Type each part of the formula as an inline or displayed formula so that you can typeset the document and check for errors.

Step 1   Let's start with $\left[\frac{n}{2}\right]$:

```
$\left[ \frac{n}{2} \right]$
```

Type this into formula.tex and test it by typesetting the document.

Step 2   Now you can do the sum:

$$\sum_{i=1}^{\left[\frac{n}{2}\right]}$$

For the superscript, you can cut and paste the formula created in Step 1 (without the dollar signs), to get

```
\[
    \sum_{i = 1}^{ \left[ \frac{n}{2} \right] }
\]
```

---

[1] The quickest way to create this file is to open mathb.tex, save it under the new name formula.tex, and delete the lines in the document environment. Then add the line
\usepackage{amssymb,amsmath}

Step 3 Next, do the two formulas in the binomial:

$$x_{i,i+1}^{i^2} \qquad \left[\frac{i+3}{3}\right]$$

Type them as separate formulas in `formula.tex`:

```
\[
    x_{i, i + 1}^{i^{2}} \qquad \left[ \frac{i + 3}{3} \right]
\]
```

Step 4 Now it's easy to do the binomial. Type the following formula by cutting and pasting the previous formulas:

```
\[
    \binom{ x_{i,i + 1}^{i^{2}} }{ \left[ \frac{i + 3}{3} \right] }
\]
```

which prints:

$$\binom{x_{i,i+1}^{i^2}}{\left[\frac{i+3}{3}\right]}$$

Step 5 Next type the formula under the square root $\mu(i)^{\frac{3}{2}}(i^2 - 1)$ as

```
$\mu(i)^{ \frac{3}{2} } (i^{2} - 1)$
```

and then the square root $\sqrt{\mu(i)^{\frac{3}{2}}(i^2 - 1)}$ as

```
$\sqrt{ \mu(i)^{ \frac{3}{2} } (i^{2} - 1) }$
```

Step 6 The two cube roots, $\sqrt[3]{\rho(i) - 2}$ and $\sqrt[3]{\rho(i) - 1}$, are easy to type:

```
$\sqrt[3]{ \rho(i) - 2 }$   $\sqrt[3]{ \rho(i) - 1 }$
```

Step 7 So now get the fraction:

$$\frac{\sqrt{\mu(i)^{\frac{3}{2}}(i^2 - 1)}}{\sqrt[3]{\rho(i) - 2} + \sqrt[3]{\rho(i) - 1}}$$

typed, cut, and pasted as

```
\[
    \frac{ \sqrt{ \mu(i)^{ \frac{3}{2}} (i^{2} -1) } }
        { \sqrt[3]{\rho(i) - 2} + \sqrt[3]{\rho(i) - 1} }
\]
```

Step 8   Finally, get the formula

$$\sum_{i=1}^{\left[\frac{n}{2}\right]} \binom{x_{i,i+1}^{i^2}}{\left[\frac{i+3}{3}\right]} \frac{\sqrt{\mu(i)^{\frac{3}{2}}(i^2-1)}}{\sqrt[3]{\rho(i)-2}+\sqrt[3]{\rho(i)-1}}$$

by cutting and pasting the pieces together, leaving only one pair of displayed math delimiters:

```
\[
   \sum_{i = 1}^{ \left[ \frac{n}{2} \right] }
      \binom{ x_{i, i + 1}^{i^{2}} }
         { \left[ \frac{i + 3}{3} \right] }
      \frac{ \sqrt{ \mu(i)^{ \frac{3}{2}} (i^{2} - 1) } }
         { \sqrt[3]{\rho(i) - 2} + \sqrt[3]{\rho(i) - 1} }
\]
```

Notice the use of

- spacing to help distinguish the braces (note that some editors help you balance the braces);
- separate lines for the various pieces.

Keep the source file readable.   Of course, this is for your benefit, since LaTeX does not care. It would also accept

```
\[\sum_{i=1}^{\left[\frac{n}{2}\right]}\binom{x_{i,i+1}^{i^{2}}}
{\left[\frac{i+3}{3}\right]}\frac{\sqrt{\mu(i)^{\frac{3}
{2}}(i^{2}-1)}}{\sqrt[3]{\rho(i)-2}+\sqrt[3]{\rho(i)-1}}\]
```

Problems arise with this haphazard style when you make a mistake. Try to find the error in the next version:

```
\[\sum_{i=1}^{\left[\frac{n}{2}\right]}\binom{x_{i,i+1}^{i^{2}}}
{\left[\frac{i+3}{3}\right]}\frac{\sqrt{\mu(i)^{\frac{3}
{2}}}(i^{2}-1)}}{\sqrt[3]{\rho(i)-2}+\sqrt[3]{\rho(i)-1}}\]
```

(Answer: \frac{3}{2} should be followed by }} and not by }}}.)

1.3   *Formula gallery*

In this section, I present the formula gallery (`gallery.tex` in the `ftp` directory), a collection of formulas—some simple, some complex—that illustrate the power of LaTeX and AMS-LaTeX. Most of the commands in these examples have not yet been discussed, but comparing the source formula with the typeset version should answer most of your questions. Occasionally, I'll give you a helping hand with some comments.

Many of these formulas are from text books and research articles. The last six are reproduced from the document `testart.tex` that was distributed by the $\mathcal{A}_{\mathcal{M}}\mathcal{S}$ with $\mathcal{A}_{\mathcal{M}}\mathcal{S}$-L*A*TEX version 1.1. Some of these examples require the amssymb and amsmath packages. So make sure to include the line

```
\usepackage{amssymb,amsmath}
```

following the `documentclass` line of any article using such constructs. The packages (if any) required for each formula shall be indicated.

Formula 1   A set-valued function:

$$x \mapsto \{\, c \in C \mid c \leq x \,\}$$

```
\[
    x \mapsto \{\, c \in C \mid c \leq x \,\}
\]
```

Note that both | and \mid print |. Use | for absolute value signs. In this formula, \mid is used because it provides extra spacing (see section 4.6.4). To equalize the spacing around $c \in C$ and $c \leq x$, a thin space was added inside each brace (see section 4.11). The same technique is used in a number of other formulas below.

Formula 2   The \left| and \right| commands print the vertical bars | whose size adjusts to the size of the formula. The \mathfrak command provides access to the *Fraktur math alphabet* (which requires the amsfonts or the eufrak package):

$$\left| \bigcup (\, I_j \mid j \in J \,) \right| < \mathfrak{m}$$

typed as

```
\[
    \left| \bigcup (\, I_{j} \mid j \in J \,) \right|
        < \mathfrak{m}
\]
```

Formula 3   Note that you need spacing both before and after the text fragment "for some" in the following example. The argument of \mbox is typeset in text mode, so a single space is recognized.

$$A = \{\, x \in X \mid x \in X_i \quad \text{for some } i \in I \,\}$$

```
\[
    A = \{\, x \in X \mid x \in X_{i}
            \mbox{\quad for some } i \in I \,\}
\]
```

Formula 4   Space to show the logical structure:

$$\langle a_1, a_2 \rangle \leq \langle a_1', a_2' \rangle \qquad \text{iff} \qquad a_1 < a_1' \quad \text{or} \quad a_1 = a_1' \text{ and } a_2 \leq a_2'$$

```
\[
    \langle a_{1}, a_{2} \rangle \leq \langle a'_{1}, a'_{2}\rangle
    \qquad \mbox{if{f}} \qquad a_{1} < a'_{1} \quad  \mbox{or}
    \quad a_{1} = a'_{1} \mbox{ and } a_{2} \leq a'_{2}
\]
```

Note that in `if{f}` (in the argument of `\mbox`) the second `f` is in braces to avoid the use of the ligature—the merging of the two f's (see section 2.4.5).

Formula 5   Here are some examples of Greek letters:

$$\Gamma_{u'} = \{ \gamma \mid \gamma < 2\chi, \; B_\alpha \nsubseteq u', \; B_\gamma \subseteq u' \}$$

```
\[
    \Gamma_{u'} = \{\, \gamma \mid \gamma < 2\chi,
    \ B_{\alpha} \nsubseteq u', \ B_{\gamma} \subseteq u' \,\}
\]
```

See Appendix A for a complete listing of Greek letters. The `\nsubseteq` command requires the amssymb package.

Formula 6   `\mathbb` gives the *Blackboard bold math alphabet* (available only in uppercase):

$$A = B^2 \times \mathbb{Z}$$

```
\[
    A = B^{2} \times \mathbb{Z}
\]
```

Blackboard bold requires the amsfonts package.

Formula 7   The `\left(` and `\right)` commands tell LaTeX to size the parentheses correctly (relative to the size of the symbols in the parentheses).

$$\left(\bigvee (s_i \mid i \in I) \right)^c = \bigwedge (s_i^c \mid i \in I)$$

```
\[
    \left( \bigvee (\, s_{i} \mid i \in I \,) \right)^{c} =
    \bigwedge (\, s_{i}^{c} \mid i \in I \,)
\]
```

Notice how the superscript is placed right on top of the subscript in s_i^c.

Formula 8

$$y \vee \bigvee ([B_\gamma] \mid \gamma \in \Gamma) \equiv z \vee \bigvee ([B_\gamma] \mid \gamma \in \Gamma) \pmod{\Phi^x}$$

```
\[
    y \vee \bigvee (\, [B_{\gamma}] \mid \gamma
    \in \Gamma \,) \equiv z \vee \bigvee (\, [B_{\gamma}]
    \mid \gamma \in \Gamma \,) \pmod{ \Phi^{x} }
\]
```

Formula 9   Use `\nolimits` so that the "limit" of the large operator is displayed as a subscript:

$$f(\mathbf{x}) = \bigvee_{\mathfrak{m}} \left(\bigwedge_{\mathfrak{m}} (x_j \mid j \in I_i) \mid i < \aleph_\alpha \right)$$

```
\[
    f(\mathbf{x}) = \bigvee\nolimits_{\!\mathfrak{m}}
    \left(\,
        \bigwedge\nolimits_{\mathfrak{m}}
      (\, x_{j} \mid j \in I_{i} \,) \mid i < \aleph_{\alpha}
    \,\right)
\]
```

The `\mathfrak` command requires the amsfonts or the eufrak package. A negative space (`\!`) was inserted to bring m a little closer to $\bigvee$ (see section 4.11).

Formula 10   The `\left.` command gives a blank left delimiter.

$$F(x)\big|_a^b = F(b) - F(a)$$

```
\[
    \left. F(x) \right|_{a}^{b} = F(b) - F(a)
\]
```

Formula 11

$$u + v \underset{\alpha}{\sim} w \overset{2}{\sim} z$$

```
\[
  u \underset{\alpha}{+} v \overset{1}{\thicksim} w
    \overset{2}{\thicksim} z
\]
```

The `\underset` and `\overset` commands require the amsmath package.

Formula 12   In this formula, \mbox would not work properly, so we use \text.

$$f(x) \overset{\text{def}}{=} x^2 - 1$$

```
\[
    f(x) \overset{ \text{def} }{=} x^{2} - 1
\]
```

This formula requires the amsmath package.

Formula 13

$$\overbrace{a + b + \cdots + z}^{n}$$

```
\[
    \overbrace{a + b + \cdots + z}^{n}
\]
```

Formula 14

$$\begin{vmatrix} a + b + c & uv \\ a + b & c + d \end{vmatrix} = 7$$

```
\[
    \begin{vmatrix}
        a + b + c & uv\\
        a + b & c + d
    \end{vmatrix}
    = 7
\]
```

$$\begin{Vmatrix} a + b + c & uv \\ a + b & c + d \end{Vmatrix} = 7$$

```
\[
    \begin{Vmatrix}
        a + b + c & uv\\
        a + b & c + d
    \end{Vmatrix}
    = 7
\]
```

The vmatrix and Vmatrix environments require the amsmath package.

Formula 15 The \mathbf{N} command makes a bold **N**. (\textbf{N} would use a different font, namely, **N**.)

$$\sum_{j\in\mathbf{N}} b_{ij}\hat{y}_j = \sum_{j\in\mathbf{N}} b_{ij}^{(\lambda)}\hat{y}_j + (b_{ii} - \lambda_i)\hat{y}_i\hat{y}$$

\[
    \sum_{j \in \mathbf{N}} b_{ij} \hat{y}_{j} =
    \sum_{j \in \mathbf{N}} b^{(\lambda)}_{ij} \hat{y}_{j} +
    (b_{ii} - \lambda_{i}) \hat{y}_{i} \hat{y}
\]

Formula 16 To produce the formula:

$$\left(\prod_{j=1}^{n} \hat{x}_j\right) H_c = \frac{1}{2}\hat{k}_{ij} \det \widehat{\mathbf{K}}(i|i)$$

try

\[
    \left(\prod^n_{\, j = 1} \hat x_{j} \right) H_{c} =
    \frac{1}{2} \hat k_{ij} \det \hat{ \mathbf{K} }(i|i)
\]

However, this produces:

$$\left(\prod_{j=1}^{n} \hat{x}_j\right) H_c = \frac{1}{2}\hat{k}_{ij} \det \hat{\mathbf{K}}(i|i)$$

Correct the overly large parentheses by using the \biggl and \biggr commands in place of \left(and \right), respectively (see section 4.6.2). Adjust the small hat over K by using \widehat:

\[
    \biggl(\prod^n_{\, j = 1} \hat x_{j} \biggr) H_{c} =
    \frac{1}{2} \hat{k}_{ij} \det \widehat{ \mathbf{K} }(i|i)
\]

Formula 17 In this formula, use \overline{I} to get $\overline{I}$ (the variant \bar{I}, which prints $\bar{I}$, is less pleasing to me):

$$\det \mathbf{K}(t = 1, t_1, \ldots, t_n) = \sum_{I\in\mathbf{n}}(-1)^{|I|} \prod_{i\in I} t_i \prod_{j\in I}(D_j + \lambda_j t_j) \det \mathbf{A}^{(\lambda)}(\overline{I}|\overline{I}) = 0$$

```
\[
    \det \mathbf{K} (t = 1, t_{1}, \dots, t_{n}) =
    \sum_{I \in \mathbf{n} }(-1)^{|I|}
    \prod_{i \in I} t_{i}
    \prod_{j \in I} (D_{j} + \lambda_{j} t_{j})
    \det \mathbf{A}^{(\lambda)} (\overline{I} | \overline{I}) = 0
\]
```

Formula 18   Note that \| provides the ‖ math symbol in this formula:

$$\lim_{(v,v') \to (0,0)} \frac{H(z+v) - H(z+v') - BH(z)(v-v')}{\|v - v'\|} = 0$$

```
\[
    \lim_{(v, v') \to (0, 0)}
    \frac{H(z + v) - H(z + v') - BH(z)(v - v')}
        {\| v - v' \|} = 0
\]
```

Formula 19   This formula uses the calligraphic math alphabet:

$$\int_{\mathcal{D}} |\overline{\partial u}|^2 \Phi_0(z) e^{\alpha|z|^2} \geq c_4 \alpha \int_{\mathcal{D}} |u|^2 \Phi_0 e^{\alpha|z|^2} + c_5 \delta^{-2} \int_A |u|^2 \Phi_0 e^{\alpha|z|^2}$$

```
\[
    \int_{\mathcal{D}} | \overline{\partial u} |^{2}
    \Phi_{0}(z) e^{\alpha |z|^2} \geq
    c_{4} \alpha \int_{\mathcal{D}} |u|^{2} \Phi_{0}
    e^{\alpha |z|^{2}} + c_{5} \delta^{-2} \int_{A}
    |u|^{2} \Phi_{0} e^{\alpha |z|^{2}}
\]
```

Formula 20   The \hdotsfor command places dots spanning multiple columns in a matrix.

The \dfrac command is the displayed variant of \frac (see section 4.4.1).

$$\mathbf{A} = \begin{pmatrix} \dfrac{\varphi \cdot X_{n,1}}{\varphi_1 \times \varepsilon_1} & (x + \varepsilon_2)^2 & \cdots & (x + \varepsilon_{n-1})^{n-1} & (x + \varepsilon_n)^n \\ \dfrac{\varphi \cdot X_{n,1}}{\varphi_2 \times \varepsilon_1} & \dfrac{\varphi \cdot X_{n,2}}{\varphi_2 \times \varepsilon_2} & \cdots & (x + \varepsilon_{n-1})^{n-1} & (x + \varepsilon_n)^n \\ \hdotsfor \\ \dfrac{\varphi \cdot X_{n,1}}{\varphi_n \times \varepsilon_1} & \dfrac{\varphi \cdot X_{n,2}}{\varphi_n \times \varepsilon_2} & \cdots & \dfrac{\varphi \cdot X_{n,n-1}}{\varphi_n \times \varepsilon_{n-1}} & \dfrac{\varphi \cdot X_{n,n}}{\varphi_n \times \varepsilon_n} \end{pmatrix} + \mathbf{I}_n$$

```
\[
    \mathbf{A} =
    \begin{pmatrix}
```

```
        \dfrac{\varphi \cdot X_{n, 1}}
              {\varphi_{1} \times \varepsilon_{1}}
      & (x + \varepsilon_{2})^{2} & \cdots
      & (x + \varepsilon_{n - 1})^{n - 1}
      & (x + \varepsilon_{n})^{n}\\
        \dfrac{\varphi \cdot X_{n, 1}}
              {\varphi_{2} \times \varepsilon_{1}}
      & \dfrac{\varphi \cdot X_{n, 2}}
              {\varphi_{2} \times \varepsilon_{2}}
      & \cdots & (x + \varepsilon_{n - 1})^{n - 1}
      & (x + \varepsilon_{n})^{n}\\
        \hdotsfor{5}\\
        \dfrac{\varphi \cdot X_{n, 1}}
              {\varphi_{n} \times \varepsilon_{1}}
      & \dfrac{\varphi \cdot X_{n, 2}}
              {\varphi_{n} \times \varepsilon_{2}}
      & \cdots & \dfrac{\varphi \cdot X_{n, n - 1}}
                       {\varphi_{n} \times \varepsilon_{n - 1}}
      & \dfrac{\varphi\cdot X_{n, n}}
              {\varphi_{n} \times \varepsilon_{n}}
    \end{pmatrix}
    + \mathbf{I}_{n}
\]
```

This formula requires the amsmath and the amssymb packages. I'll show in section 9.1.2 how to write this formula so that it's short and more readable.

1.4 Typing equations and aligned formulas

1.4.1 Equations

The equation environment creates a displayed math formula and automatically generates a number. The equation

(1)
$$\int_0^\pi \sin x \, dx = 2$$

is typed as

```
\begin{equation} \label{E:firstInt}
   \int_{0}^{\pi} \sin x \, dx = 2
\end{equation}
```

Of course, the number generated depends on how many equations precede the given one.

To refer to this formula without having to remember a (changeable) number, assign a *name* to the equation in the argument of a \label command; I'll call the name of the equation a *label*. In this section, let's call the first equation "firstInt" (first integral). I use the convention that the label of an equation starts with "E:".

The number of this formula is referenced with the \ref command. For example, to get the reference "see (1)", type

```
see~(\ref{E:firstInt})
```

Alternatively, with the amsmath package, you can use the \eqref command. For instance,

```
see~\eqref{E:firstInt}
```

also produces "see (1)".

An advantage of this cross-referencing system is that if a new equation is introduced, or the existing equations are rearranged, the numbering will automatically be adjusted to reflect these changes.

Rule ■ Typeset twice

For renumbering to work, you have to typeset the source file twice.

See sections 6.3.2 and E.2.4. LaTeX will send a warning if you forget.

At the end of the typesetting, LaTeX stores the labels in the aux file (see section 1.11.3). For every label, it stores the number the label is associated with and also the page number on which the label occurs in the typeset version.

An equation will be numbered whether or not there is a \label command attached to it. Of course, if there is no \label command, the number generated by LaTeX for the equation can't be referenced automatically.

The system described here is called *symbolic referencing*. The argument of \label is the "symbol" for the number, and \ref provides the referencing. LaTeX uses the same mechanism for all numberings it automatically generates: numbering of section titles, equations, theorems, lemmas, and bibliographic references—except that for bibliographic references the commands are \bibitem and \cite, respectively (see section 1.7.4).

With the amsmath package, equations can also be *tagged* by attaching a name to the formula with the \tag command; the tag replaces the number.

Example:

$$(\text{Int}) \qquad\qquad \int_0^\pi \sin x \, dx = 2$$

is typed as

```
\begin{equation}
    \int_{0}^{\pi} \sin x \, dx = 2 \tag{Int}
\end{equation}
```

Tags (of the type discussed here) are **absolute**; this equation is always referred to as (Int). Equation numbers, on the other hand, are **relative**; they change as equations are added, deleted, or rearranged.

1.4.2   Aligned formulas

LaTeX, with the help of the amsmath package, has many ways to typeset multiline formulas. Right now, you'll be introduced to three constructs: *simple align, annotated align*, and *cases*; see Chapter 5 for a discussion of many others.

The align math environment is used for simple and annotated align. *Each line* in this environment is an equation, which LaTeX automatically numbers.

Simple align

Simple align is used to align two or more formulas. To obtain the formulas

$$(2) \qquad\qquad r^2 = s^2 + t^2$$

$$(3) \qquad\qquad 2u + 1 = v + w^\alpha$$

$$(4) \qquad\qquad x = \frac{y + z}{\sqrt{s + 2u}}$$

type (using \\ as a line separator)

```
\begin{align}
    r^{2}  &= s^{2} + t^{2}                  \label{E:eqn1}\\
    2u + 1 &= v + w^{\alpha}                 \label{E:eqn2}\\
    x      &= \frac{y + z}{\sqrt{s + 2u}}    \label{E:eqn3}
\end{align}
```

(These equations are numbered (2), (3), and (4) because they are preceded by one numbered equation earlier in this section.)

The align environment can also be used to break a long formula into two. Since numbering both lines is undesirable, you may prevent the numbering of the second line with the \notag command.

$$(5) \qquad h(x) = \int \left(\frac{f(x) + g(x)}{1 + f^2(x)} + \frac{1 + f(x)g(x)}{\sqrt{1 - \sin x}} \right) dx$$

$$= \int \frac{1 + f(x)}{1 + g(x)}\, dx - 2 \tan^{-1}(x - 2)$$

This formula may be typed as

```
\begin{align} \label{E:longInt}
   h(x) &= \int
    \left(
          \frac{ f(x) + g(x) }
               { 1+ f^{2}(x) }
          + \frac{ 1+ f(x)g(x) }
                 { \sqrt{1 - \sin x} }
    \right) \, dx\\
          &= \int \frac{ 1 + f(x) }
                       { 1 + g(x) }
          \, dx - 2 \tan^{-1}(x-2) \notag
\end{align}
```

See the `split` subsidiary math environment in section 5.5.2 for a better way to split a long formula into (two or more) aligned parts, and on how to center the formula number (5) between the two lines.

The rules are easy for simple align:

Rule ■ Simple align

- Separate the lines with \\.
- In each line, indicate the alignment point with &.
- Place a \notag in each line that you do not wish numbered.
- Place a \label in each numbered line you may want to reference with \ref or \eqref.

Annotated align

Annotated align will align the formulas and the annotation (explanatory text) separately:

$$(6) \qquad x = x \wedge (y \vee z) \qquad\qquad \text{(by distributivity)}$$
$$= (x \wedge y) \vee (x \wedge z) \qquad\qquad \text{(by condition (M))}$$
$$= y \vee z.$$

This is typed as:

```
\begin{align} \label{E:DoAlign}
   x &= x \wedge (y \vee z)
                        & &\text{(by distributivity)}\\
      &= (x \wedge y) \vee (x \wedge z)
                        & &\text{(by condition (M))} \notag\\
      &= y \vee z. \notag
\end{align}
```

The rules for annotated align are similar to the rules of simple align. In each line, in addition to the alignment point (marked by &), there is also a mark for the start of the annotation: & &.

The `align` environment does much more than simple and annotated aligns (see section 5.4).

Cases

The `cases` construct is a *subsidiary math environment*; it must be used in a displayed math environment or in an `equation` environment (see section 5.5). Here is a typical example:

$$f(x) = \begin{cases} -x^2, & \text{if } x < 0; \\ \alpha + x, & \text{if } 0 \leq x \leq 1; \\ x^2, & \text{otherwise.} \end{cases}$$

which may be typed as follows:

```
\[
    f(x)=
    \begin{cases}
        -x^{2},        &\text{if $x < 0$;}\\
        \alpha + x,    &\text{if $0 \leq x \leq 1$;}\\
        x^{2},         &\text{otherwise.}
    \end{cases}
\]
```

The rules for cases are simple:

Rule ■ cases

- Separate the lines with \\.
- In each line, indicate the alignment point for the annotation with &.

1.5 *The anatomy of an article*

The `sampart.tex` sample article (typeset on pages 361–363) uses the $\mathcal{AMS}$ article document class, amsart. In this introductory chapter, I want to start off with the popular `article` document class of LaTeX, which is easier to use. So we'll use a simplified and shortened sample article, `intrart.tex` (in the `ftp` directory). Type it in as we discuss the parts of an article.

The *preamble* of an article is the initial part of the source file up to the line

```
\begin{document}
```

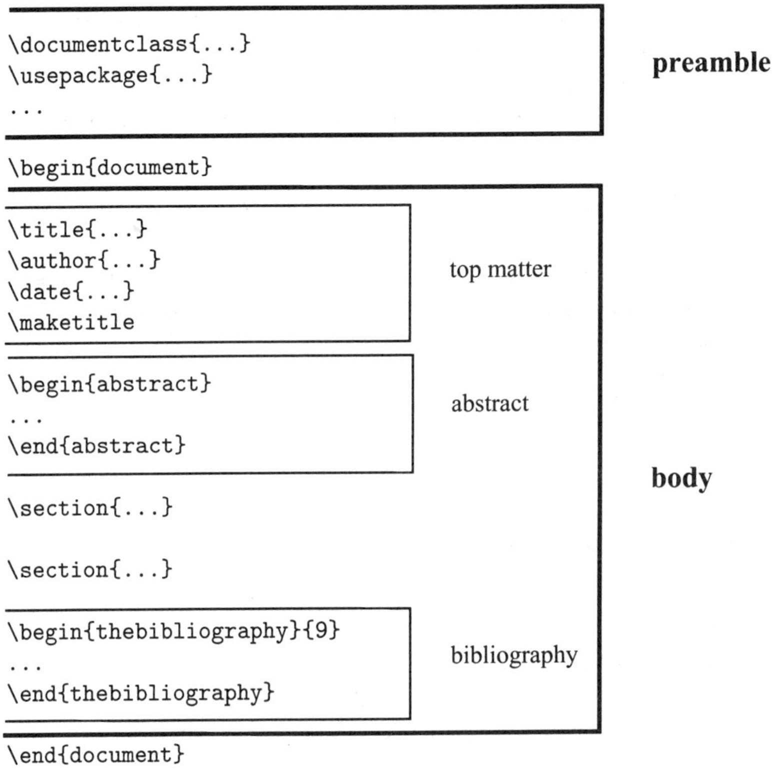

```
\documentclass{...}
\usepackage{...}
...
```
                                                                    preamble

```
\begin{document}
```

```
\title{...}
\author{...}
\date{...}
\maketitle
```
                                    top matter

```
\begin{abstract}
...
\end{abstract}
```
                                    abstract
                                                                    body

```
\section{...}
```

```
\section{...}
```

```
\begin{thebibliography}{9}
...
\end{thebibliography}
```
                                    bibliography

```
\end{document}
```

Figure 1.1: A schematic view of an article

See Figure 1.1. The preamble contains instructions for the entire article, for instance, the

```
\documentclass
```

command.

Here is the preamble of the introductory sample article:

```
% Introductory sample article: intrart.tex
% Typeset with LaTeX format

\documentclass{article}
\usepackage{amssymb,amsmath}
\newtheorem{theorem}{Theorem}
\newtheorem{definition}{Definition}
\newtheorem{notation}{Notation}
```

The preamble names the document class, `article`, and then names the LaTeX enhancements, or packages, used by the article. This article loads two packages:

The amssymb package provides the names of all the math symbols in Appendix A and the amsmath package provides many of the math constructs used.

A *proclamation* is a theorem, definition, corollary, note, and so on. In the preamble, three proclamations are defined. For instance,

```
\newtheorem{theorem}{Theorem}
```

defines the theorem environment, which you can use in the body of your article (see section 1.7.3). LaTeX will automatically number and visually format the theorems.

The article proper, called the *body* of the article, is contained in the document environment, that is, between the lines

```
\begin{document}
```

and

```
\end{document}
```

as illustrated in Figure 1.1. The body of the article is also logically split up into several parts; we'll discuss these in detail in section 6.1.

The body of the article starts with the *top matter*, which contains the "title page" information. It follows the line:

```
\begin{document}
```

and concludes with the line

```
\maketitle
```

Here is the top matter of the introductory sample article:

```
\title{A construction of complete-simple\\
       distributive lattices}
\author{George~A. Menuhin\thanks{Research supported
   by the NSF under grant number~23466.}\\
   Computer Science Department\\
   University of Winnebago\\
   Winnebago, Minnesota 23714\\
   \texttt{menuhin@ccw.uwinnebago.edu}}
\date{March 15, 1995}
\maketitle
```

The body continues with an (optional) abstract, contained in an abstract environment:

```
\begin{abstract}
   In this note we prove that there exist \emph{complete-simple
   distributive lattices}, that is, complete distributive
   lattices in which there are only two complete congruences.
\end{abstract}
```

And here is the rest of the body of the introductory sample article:

```
\section{Introduction} \label{S:intro}
In this note we prove the following result:

\begin{theorem}
    There exists an infinite complete distributive lattice $K$
    with only the two trivial complete congruence relations.
\end{theorem}

\section{The $\Pi^{*}$ construction} \label{S:P*}
The following construction is crucial in our proof of our Theorem:

\begin{definition} \label{D:P*}
    Let $D_{i}$, $i \in I$, be complete distributive
    lattices satisfying condition~\textup{(J)}.  Their
    $\Pi^{*}$ product is defined as follows:
    \[
       \Pi^{*} ( D_{i} \mid i \in I ) =
         \Pi ( D_{i}^{-} \mid i \in I ) + 1;
    \]
    that is, $\Pi^{*} ( D_{i} \mid i \in I )$ is
    $\Pi ( D_{i}^{-} \mid i \in I )$ with a new unit element.
\end{definition}

\begin{notation}
    If $i \in I$ and $d \in D_{i}^{-}$, then
    \[
        \langle \dots, 0, \dots, \overset{i}{d}, \dots, 0,
          \dots \rangle
    \]
    is the element of $\Pi^{*} ( D_{i} \mid i \in I )$ whose
    $i$th component is $d$ and all the other components
    are $0$.
\end{notation}

See also Ernest~T. Moynahan~\cite{eM57a}.


Next we verify the following result:

\begin{theorem} \label{T:P*}
    Let $D_{i}$, $i \in I$, be complete distributive
    lattices satisfying condition~\textup{(J)}.  Let $\Theta$
    be a complete congruence relation on
```

```
    $\Pi^{*} ( D_{i} \mid i \in I )$.
    If there exists an $i \in I$ and a $d \in D_{i}$ with
    $d < 1_{i}$ such that for all $d \leq c < 1_{i}$,
    \begin{equation} \label{E:cong1}
       \langle \dots, 0, \dots,\overset{i}{d},
       \dots, 0, \dots \rangle \equiv \langle \dots, 0, \dots,
       \overset{i}{c}, \dots, 0, \dots \rangle \pmod{\Theta},
    \end{equation}
    then $\Theta = \iota$.
\end{theorem}


\emph{Proof.} Since
\begin{equation} \label{E:cong2}
    \langle \dots, 0, \dots, \overset{i}{d}, \dots, 0,
       \dots \rangle \equiv \langle \dots, 0, \dots,
       \overset{i}{c}, \dots, 0, \dots \rangle \pmod{\Theta},
\end{equation}
and $\Theta$ is a complete congruence relation, it follows
from condition~(C) that
\begin{align} \label{E:cong}
   & \langle \dots, \overset{i}{d}, \dots, 0,
    \dots \rangle \equiv\\
   &\qquad \qquad \quad \bigvee ( \langle \dots, 0, \dots,
   \overset{i}{c}, \dots, 0, \dots \rangle \mid d \leq c < 1 )
   \equiv 1 \pmod{\Theta}. \notag
\end{align}


Let $j \in I$, $j \neq i$, and let $a \in D_{j}^{-}$.
Meeting both sides of the congruence \eqref{E:cong2} with
$\langle \dots, 0, \dots, \overset{j}{a}, \dots, 0,
\dots \rangle$, we obtain

\begin{align} \label{E:comp}
   0 = & \langle \dots, 0, \dots, \overset{i}{d}, \dots, 0, \dots
     \rangle \wedge \langle \dots, 0, \dots, \overset{j}{a},
     \dots, 0, \dots \rangle \equiv\\
     &\langle \dots, 0, \dots, \overset{j}{a}, \dots, 0, \dots
     \rangle \pmod{\Theta}, \notag
\end{align}
Using the completeness of $\Theta$ and \eqref{E:comp},
we get:
\[
```

```
        0 \equiv \bigvee ( \langle \dots, 0, \dots, \overset{j}{a},
        \dots, 0,\dots\rangle \mid a \in D_{j}^{-} ) = 1 \pmod{\Theta},
\]
hence $\Theta = \iota$.

\begin{thebibliography}{9}
    \bibitem{sF90}
        Soo-Key Foo, \emph{Lattice constructions}, Ph.D. thesis,
        University of Winnebago, Winnebago MN, December 1990.
    \bibitem{gM68}
        George~A. Menuhin, \emph{Universal Algebra}, D.~van Nostrand,
        Princeton-Toronto-London-Mel\-bourne, 1968.
    \bibitem{eM57}
        Ernest~T. Moynahan, \emph{On a problem of M.~H. Stone}, Acta
        Math. Acad. Sci. Hungar. \textbf{8} (1957), 455--460.
    \bibitem{eM57a}
        Ernest~T. Moynahan, \emph{Ideals and congruence relations in
        lattices. II}, Magyar Tud. Akad. Mat. Fiz. Oszt. K\"{o}zl.
        \textbf{9} (1957), 417--434.
\end{thebibliography}
```

At the end of the body, the *bibliography* is typed between the lines

```
\begin{thebibliography}{9}
\end{thebibliography}
```

The argument "9" of the thebibliography environment tells LaTeX to make room for single digit numbering, since in this article there are fewer than 10 articles. In the typeset article, the bibliography is entitled "References".

Observe that we refer to condition (J) in the definition as \textup{(J)}. We do this so that if the text of the definition is emphasized (as it is), then (J) should still be typeset as (J) and not as *(J)*; see section 2.6.4 for the \textup command.

1.5.1   *The typeset article*

Here is the typeset introductory sample article (note that the equation numbers are on the right, the default in the article document class; elsewhere in this book you find the $\mathcal{AMS}$ default, equations on the left—see sections 7.1.2 and 8.4 on how to change the default).

A construction of complete-simple distributive lattices

George A. Menuhin[*]
Computer Science Department
Winnebago, Minnesota 23714
menuhin@ccw.uwinnebago.edu

March 15, 1995

Abstract

In this note we prove that there exist *complete-simple distributive lattices*, that is, complete distributive lattices in which there are only two complete congruences.

1 Introduction

In this note we prove the following result:

Theorem 1 *There exists an infinite complete distributive lattice K with only the two trivial complete congruence relations.*

2 The Π^* construction

The following construction is crucial in our proof of our Theorem:

Definition 1 *Let D_i, $i \in I$, be complete distributive lattices satisfying condition (J). Their Π^* product is defined as follows:*

$$\Pi^*(D_i \mid i \in I) = \Pi(D_i^- \mid i \in I) + 1;$$

that is, $\Pi^*(D_i \mid i \in I)$ is $\Pi(D_i^- \mid i \in I)$ with a new unit element.

Notation 1 *If $i \in I$ and $d \in D_i^-$, then*

$$\langle \ldots, 0, \ldots, \overset{i}{d}, \ldots, 0, \ldots \rangle$$

is the element of $\Pi^*(D_i \mid i \in I)$ whose ith component is d and all the other components are 0.

---

[*]Research supported by the NSF under grant number 23466.

1

See also Ernest T. Moynahan [4].

Next we verify the following result:

Theorem 2 *Let D_i, $i \in I$, be complete distributive lattices satisfying condition* (J). *Let Θ be a complete congruence relation on $\Pi^*(D_i \mid i \in I)$. If there exists an $i \in I$ and a $d \in D_i$ with $d < 1_i$ such that for all $d \leq c < 1_i$,*

$$\langle \ldots, 0, \ldots, \overset{i}{d}, \ldots, 0, \ldots \rangle \equiv \langle \ldots, 0, \ldots, \overset{i}{c}, \ldots, 0, \ldots \rangle \pmod{\Theta}, \qquad (1)$$

then $\Theta = \iota$.

Proof. Since

$$\langle \ldots, 0, \ldots, \overset{i}{d}, \ldots, 0, \ldots \rangle \equiv \langle \ldots, 0, \ldots, \overset{i}{c}, \ldots, 0, \ldots \rangle \pmod{\Theta}, \qquad (2)$$

and Θ is a complete congruence relation, it follows from condition (C) that

$$\langle \ldots, \overset{i}{d}, \ldots, 0, \ldots \rangle \equiv \qquad\qquad\qquad\qquad\qquad (3)$$
$$\bigvee (\langle \ldots, 0, \ldots, \overset{i}{c}, \ldots, 0, \ldots \rangle \mid d \leq c < 1) \equiv 1 \pmod{\Theta}.$$

Let $j \in I$, $j \neq i$, and let $a \in D_j^-$. Meeting both sides of the congruence (2) with $\langle \ldots, 0, \ldots, \overset{j}{a}, \ldots, 0, \ldots \rangle$, we obtain

$$0 = \langle \ldots, 0, \ldots, \overset{i}{d}, \ldots, 0, \ldots \rangle \wedge \langle \ldots, 0, \ldots, \overset{j}{a}, \ldots, 0, \ldots \rangle \equiv \qquad (4)$$
$$\langle \ldots, 0, \ldots, \overset{j}{a}, \ldots, 0, \ldots \rangle \pmod{\Theta},$$

Using the completeness of Θ and (4), we get:

$$0 \equiv \bigvee (\langle \ldots, 0, \ldots, \overset{j}{a}, \ldots, 0, \ldots \rangle \mid a \in D_j^-) = 1 \pmod{\Theta},$$

hence $\Theta = \iota$.

References

[1] Soo-Key Foo, *Lattice Constructions*, Ph.D. thesis, University of Winnebago, Winnebago, MN, December 1990.

[2] George A. Menuhin, *Universal Algebra*, D. van Nostrand, Princeton-Toronto-London-Melbourne, 1968.

[3] Ernest T. Moynahan, *On a problem of M. H. Stone*, Acta Math. Acad. Sci. Hungar. **8** (1957), 455–460.

[4] Ernest T. Moynahan, *Ideals and congruence relations in lattices. II*, Magyar Tud. Akad. Mat. Fiz. Oszt. Közl. **9** (1957), 417–434.

1.6 *Article templates*

Before you start writing your first article, I suggest you create two article templates for your own use.

There are two templates for articles written in the `article` document class in this book: `article.tpl` for articles with one author and `article2.tpl` for articles with two authors.[2] You can find these in the `ftp` directory (see page 4). So copy `article.tpl` into the work directory or type it in as follows:

```
% Sample file: article.tpl
% Typeset with LaTeX format

\documentclass{article}
\usepackage{amsmath,amssymb}

\newtheorem{theorem}{Theorem}
\newtheorem{lemma}{Lemma}
\newtheorem{proposition}{Proposition}
\newtheorem{definition}{Definition}
\newtheorem{corollary}{Corollary}
\newtheorem{notation}{Notation}

\begin{document}
\title{%
   titleline1\\
   titleline2}
\author{name\thanks{support}\\
   addressline1\\
   addressline2\\
   addressline3}
\date{date}
\maketitle

\begin{abstract}
   abstract
\end{abstract}

\begin{thebibliography}{99}

\end{thebibliography}
\end{document}
```

---

[2]In section 8.3, we discuss a template file, `amsart.tpl`, for the $\mathcal{AMS}$ document class `amsart`.

Now copy `article2.tpl` into the work directory, or type it in. It is identical to `article.tpl` except for the argument of the `\author` command:

```
\author{name1\thanks{support1}\\
    address1line1\\
    address1line2\\
    address1line3
    \and
    name2\thanks{support2}\\
    address2line1\\
    address2line2\\
    address2line3}
```

Note the `\and` command; it separates the two authors.

Now let's customize the template files. Open `article.tpl` and save it under a name of your choosing; I saved it under the name `ggart.tpl` (in the `ftp` directory—see page 4). In this personalized template file, I edit the top matter:

```
\title{titleline1\\
        titleline2}
\author{G. Gr\"{a}tzer\thanks{Research supported by the
                        NSERC of Canada.}\\
    University of Manitoba\\
    Department of Mathematics\\
    Winnipeg, Man. R3T 2N2\\
    Canada}
\date{date}
```

I did not edit the `\title` lines because they change from article to article. There is also a personalized `ggart2.tpl` for two authors.

1.7 Your first article

Your first article will be typeset using the `article` document class. To start, open the personalized article template created in section 1.6, and save it under the name of your first article. The name must be **one word** (no spaces) ending with `.tex`.

1.7.1 Editing the top matter

Edit the top matter to contain the article information (title, date, and so on). Here are some simple rules to follow:

Rule ■ Top matter for the `article` document class

1. If the title is only one line long, then there is no \\ in the argument of the \title command; otherwise, separate the lines of the title with \\. There is no \\ at the end of the last line.
2. Separate the lines of the address with \\. There is no \\ at the end of the last line.
3. \thanks places a footnote at the bottom of the first page. If it is not needed, delete it.
4. Multiple authors are separated by \and. There is only one \author command, and it contains all the information (name, address, support) about all the authors.
5. The \title command is the only compulsory command. The others are optional.
6. If there is no \date command, the current date will be shown. If you do not want a date, type the form \date{}; if you want a specific date, say February 21, 1995, write
\date{February 21, 1995}

1.7.2 Sectioning

An article, as a rule, is divided into sections. To start the section entitled "Introduction", type

\section{Introduction} \label{S:intro}

Introduction is the title of the section, S:intro is the label. I use the convention that "S:" starts the label for a section. The number of the section is automatically assigned by LaTeX, and you can refer to this section number by \ref{S:intro}, as in

In section~\ref{S:intro}, we introduce ...

(the tilde ˜ is an unbreakable space, it keeps the word "section" and the section number together—see section 2.4.3).

For instance, the section title of this section was typed as follows:

\section{Typing your first article} \label{S:FirstArticle}

A reference to this section is made by typing

```
\ref{S:FirstArticle}
```

Sections have subsections, and subsections have subsubsections, followed by paragraphs and subparagraphs. The corresponding commands are

```
\subsection    \subsubsection   \paragraph   \subparagraph
```

1.7.3   Invoking proclamations

In the preamble of `article.tpl`, you typed the theorem, lemma, proposition, definition, corollary, and notation proclamations. Each of these proclamations defines an environment. For example, type a theorem in a `theorem` environment; the body of the theorem (that is, the part of the source file that produces the theorem) is between the two lines:

```
\begin{theorem} \label{T:xxx}
```

and

```
\end{theorem}
```

where `T:xxx` is the label for the theorem. Of course, xxx should be somewhat descriptive of the contents of the theorem. The theorem number is automatically assigned by LaTeX, and it can be referenced by `\ref{T:xxx}` as in

```
it follows from Theorem~\ref{T:xxx}
```

(the tilde ~ keeps the word "Theorem" and the theorem number together—see section 2.4.3). I use the convention that the label for a theorem starts with "T:".

1.7.4   Inserting references

Finally, we discuss the bibliography. Below are typical entries for the most often used types of references: an article in a journal, a book, an article in a conference proceedings, an article (chapter) in a book, a Ph.D. thesis, and a technical report (see `inbibl.tpl` in the `ftp` directory).

```
\bibitem{eM57}
   Ernest~T. Moynahan, \emph{On a problem of M.~H. Stone},
      Acta Math. Acad. Sci. Hungar. \textbf{8} (1957), 455--460.

\bibitem{gM68}
   George~A. Menuhin, \emph{Universal Algebra}, D.~van Nostrand,
      Princeton-Toronto-London-Melbourne, 1968.

\bibitem{pK69}
```

```
        Peter~A. Konig, \emph{Composition of functions}, Proceedings of
        the Conference on Universal Algebra (Kingston, 1969).

\bibitem{hA70}
        Henry~H. Albert, \emph{Free torsoids}, Current Trends in
        Lattice Theory, D.~van Nostrand, 1970.

\bibitem{sF90}
        Soo-Key Foo, \emph{Lattice constructions}, Ph.D. thesis,
        University of Winnebago, 1990.

\bibitem{gF86}
        Grant~H. Foster, \emph{Computational complexity in lattice
        theory}, Tech. report, Carnegie Mellon University, 1986.
```

I use the convention that the label for the \bibitem consists of the initials of the author and the year of publication: a publication by Andrew B. Reich in 1987 would have the label aR87 (the second publication would be aR87a). For joint publications, the label consists of the initials of the authors and the year of publication; for instance, a publication by John Bradford and Andrew B. Reich in 1987 would have the label BR87. Of course, you can use any label you choose (subject to the rule in section 6.4.2).

Suppose you want to include as the fifth item in the bibliography the following article:

John Bradford and Andrew B. Reich, *Duplexes in posets*, Proc. Amer. Math. Soc. **112** (1987), 115–125.

Modeling it after Moynahan's article, type it as:

```
\bibitem{BR87}
        John~Bradford and Andrew~B. Reich, \emph{Duplexes in posets},
        Proc. Amer. Math. Soc. \textbf{112} (1987), 115--125.
```

A reference to this article is made with \cite{BR87}, for instance:

this result was first published in [5]

typed as

```
this result was first published in~\cite{BR87}
```

Note that you have to arrange the references in the thebibliography environment in the order you wish to see them. LaTeX only takes care of the numbering and the citations in the text.

Tip   The `thebibliography` environment properly handles periods. You do not have to mark periods for abbreviations (in the form `.\`␣—as discussed in section 2.2.2) in the name of a journal, so

```
Acta Math. Acad. Sci. Hungar.
```

is correct.

1.8   LaTeX error messages

You'll probably make a number of mistakes in your first article. The mistakes come in various forms:

- Typographical errors, which LaTeX will blindly typeset. View the typeset version, find the errors, and correct the source file.
- Errors in mathematical formulas or in the formatting of the text.
- Errors in your instructions—commands and environments—to LaTeX.

Let's look at some examples by introducing a number of errors in the source file of the `intrart.tex` introductory sample article and see what error messages occur.

Example 1   Go to line 21 (you do not have to count lines, since most editors have a "go to line" command) and remove the closing brace so it reads:

```
\begin{abstract
```

Upon typesetting `intrart.tex`, LaTeX informs you of a mistake:

```
Runaway argument?
{abstract
! Paragraph ended before \end was complete.
<to be read again>
                    \par
1.26
```

Line 26 of the file is the line after `\end{abstract}`. From the error message, you can tell that something is wrong with the `abstract` environment.

Example 2   Now correct line 21, go to line 25, change it from

```
\end{abstract}
```

to

```
\end{abstrac}
```

and typeset again. LaTeX will inform you:

```
! LaTeX Error: \begin{abstract} on input line 21
ended by \end{abstrac}.
```

```
l.25 \end{abstrac}
```

Pressing return, LaTeX will recover from this error.

Example 3   Instead of correcting the error in line 25, comment it out:

```
% \end{abstrac}
```

and introduce an additional error in line 67. This line presently reads:

```
    lattices satisfying condition~\textup{(J)}.  Let $\Theta$
```

Change \Theta to \Teta:

```
    lattices satisfying condition~\textup{(J)}.  Let $\Teta$
```

Typesetting the article now, the message is:

```
! Undefined control sequence.
l.67 ...xtup{(J)}.  Let $\Teta
                              $
```

and pressing return gives the message:

```
! LaTeX Error: \begin{abstract} on input line 21 ended
by \end{document}.
```

```
l.131 \end{document}
```

These two mistakes are easy to identify. \Teta is a typo for \Theta. Observe how LaTeX tries to match

```
\begin{abstract}
```

with

```
\end{document}
```

Undo the two changes (lines 25 and 67).

Example 4   In line 73, change

```
    \langle \dots, 0, \dots,\overset{i}{d},
```

to

```
    \langle \dots, 0, \dots,\overset{i}{d,
```

This results in the message:

```
Runaway argument?
\def \\{\@amsmath@err {\Invalid@@ \\}\@eha } \label {E\ETC.
! Paragraph ended before \equation was complete.
<to be read again>
                    \par
1.79
```

Line 79 is the blank line following \end{theorem}. LaTeX skipped over the defective construct \overset and the incomplete equation environment, indicating the error past the end of the theorem environment. The error message indicates that the error may have been caused by the new paragraph (\par). Of course, there can be no new paragraph in either the second argument of \overset or the displayed formula. The solution does not come easily except by isolating the last paragraph and investigating it.

Error messages from LaTeX are not always as helpful as one would like, but there is always some information to be gleaned from them. As a rule, the error message should at least inform you of the line number (or paragraph or formula) where the error was caught. Try to identify the structure that caused the error: a command, an environment, or so forth. Keep in mind that it could be quite far from the line where LaTeX indicated the error. Try reading the section of this book that describes that command or environment; it should help in correcting the error.

The next best defense is to isolate your problem. Create a current.tex file that is the same as the present article, except that there is only one paragraph in the document environment. When this paragraph is typeset correctly, cut and paste it into your source file. If there is only one paragraph in the document, the error is easier to find. If the error is of the type as in the last example, split the paragraph into smaller paragraphs. See also section 2.5 on how to use the comment environment for finding errors.

1.9   *Logical and visual design*

This book attempts to show how to typeset an article, **not** how to write it. Nevertheless, it seems appropriate to point out some approaches to article design.

The typeset version of our intrart.tex introductory sample article (pp. 39–40) looks impressive. (For another example of a typeset article, see sampart.tex on pp. 361–363.) To produce an article like this, you have to realize that there are two aspects of article design: the *visual* and the *logical*. Let's borrow an example from the sample article to illustrate this: a theorem. You tell LaTeX to typeset a theorem and number it. Here is how you type the theorem:

```
\begin{theorem} \label{T:P*}
```

```
Let $D_{i}$, $i \in I$, be complete distributive
lattices satisfying condition~\textup{(2)}.  Let $\Theta$
be a complete congruence relation on
$\Pi^{*} ( D_{i} \mid i \in I )$.
If there exists an $i \in I$ and a $d \in D_{i}$ with
$d < 1_{i}$ such that for all $d \leq c < 1_{i}$,
\begin{equation} \label{E:cong1}
   \langle \dots, 0, \dots,\overset{i}{d},
   \dots, 0, \dots \rangle \equiv \langle \dots, 0, \dots,
   \overset{i}{c}, \dots, 0, \dots \rangle \pmod{\Theta},
\end{equation}
then $\Theta = \iota$.
\end{theorem}
```

You find the typeset form on page 40.

The logical design is the theorem itself, which is placed in the theorem environment. For the visual design, LaTeX makes literally hundreds of decisions: the vertical space before and after the theorem; the bold **Theorem** heading and its numbering; the vertical space before and after the equation, and its numbering; the spacing of all the math symbols (inline and displayed formulas are spaced differently); the text of the theorem to be emphasized; and so on.

The decisions were made by professional designers, whose expertise is hidden in TeX itself, in LaTeX, in the document class, and in the packages. Could you have typeset this theorem yourself? Probably not. Aesthetic decisions are difficult for lay people to make. But even if you could have guessed the correct spacing, you would have faced the problem of consistency (guaranteeing that the next theorem will look the same), and just as importantly, you would have spent a great deal of time and energy on the *visual design* of the theorem, as opposed to the *logical design*. The idea is to concentrate on the logical design and let LaTeX take care of the visual design.

This approach has the advantage that by changing the document class (or its options; see sections 7.1.2 and 8.4), the visual design can be changed. If you code the visual design into the article ("hard coding" it, as a programmer would say), it's very difficult to change.

LaTeX uses four major tools to separate the logical and visual designs of an article:

Commands Information is given to LaTeX as arguments of commands; it's up to LaTeX to process the information. For instance, the title page information (especially in the amsart document class) is given in this form; the organization of the title page is completely up to the document class and its options.

A more subtle example is the use of a command for distinguishing a term or notation. For instance, you may want to use an \env command for environment names. You may define \env as follows (\newcommand is explained in section 9.1.1):

```
\newcommand{\env}[1]{\texttt{#1}}
```

which typesets all environment names in typewriter style (see section 2.6.2). Logically, you have decided that an environment name should be marked up. Visually, you may change the decision any time. By changing the definition to

```
\newcommand{\env}[1]{\textbf{#1}}
```

all environment names will be typeset in bold (see section 2.6.7).

The following more mathematical example is taken from `sampart2.tex` (see Appendix D and the `ftp` directory). This article defines the construct $D^{\langle 2 \rangle}$ with the command

```
\newcommand{\Ds}{D^{\langle 2 \rangle}}
```

If a referee (or coauthor) suggests a different notation, changing this *one line* will carry out the change throughout the whole article.

Environments Important logical structures are placed in environments. For instance, you can give a list as an environment by saying that this is a list and these are the items (see section 3.1). Again, exactly how the list is typeset is up to LaTeX; you can even switch from one list type to another by just changing the name of the environment.

Proclamations These define numbered environments. If the amsthm package is used, you can further specify which one of three styles to use for typesetting; at any time you can change the style or the numbering scheme in the preamble (see the typeset `sampart.tex` on pages 361–363 for examples of proclamations printed in the three styles).

Cross-referencing Since theorems and sections are logical units, they can be freely moved around. This gives tremendous freedom in reorganizing the source file to improve the logical design.

You write articles to communicate. The closer you get to a separation of logical and visual design, the more you are able to concentrate on communicating your ideas. Of course, you can never quite reach this ideal. For instance, a "line too wide" warning (see sections 1.1.3 and 2.7.1) is a problem of visual design. When the journal changes the document class, unless the new document class retains the same fonts and line width, new "line too wide" problems arise. However, LaTeX is successful well over 95% of the time in solving visual design problems without your intervention. This is getting fairly close to the ideal.

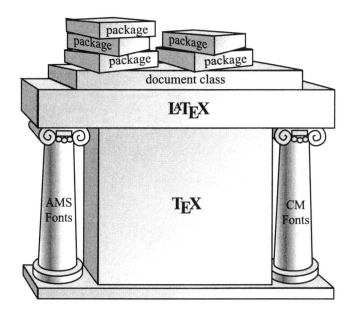

Figure 1.2: The structure of LaTeX

1.10   A brief overview

Having finished the short course, maybe it's time to pause and get a brief overview of how LaTeX works. As I pointed out in the Introduction, at the core of LaTeX is a *programming language* called TeX, providing many typesetting instructions. Along with TeX comes a set of fonts called *Computer Modern* (CM). The CM fonts and the TeX programming language form the foundation of a typical TeX system.

TeX is expandable, that is, additional commands can be defined in terms of more basic ones. One of the best known expansions of TeX is LaTeX; it introduces the idea of a *logical unit* that you read about in section 1.9.

Visual layout in LaTeX is determined by the *document class*; for example, you now have some familiarity with the `article` document class. Expansions of LaTeX are called *packages*; you have already come across the amssymb and amsmath packages.

The structure of LaTeX is illustrated in Figure 1.2. This figure suggests that, in order to work with a LaTeX document, you first have to install TeX and the CM fonts, then LaTeX, and finally specify the document class and the necessary packages. The AMSFonts font set is useful but it's not absolutely necessary.

Figure 1.2 illustrates my view of TEX and LATEX: it is the foundation on which many useful packages—extensions of LATEX—are built. It is essential that you understand the packages that make your work easier. An important example of this is the central focus of this book: typesetting math in LATEX. When typesetting math, invoke the amsmath package. In Part I, you invoke the amsmath package directly; in later parts of this book, I point out when a described feature needs the amsmath (or some other) package. The $\mathcal{AMS}$ document classes automatically load the amsmath and amsfonts packages.

1.11   *Using LATEX*

Figure 1.3 illustrates the steps taken to produce a typeset document. As illustrated in Figure 1.3, you open the source file or create a new one using an editor; call the source file `myart.tex`. Once the document is ready, typeset it with TEX using the LaTeX format. This step produces three files:

- `myart.dvi`, the typeset article in machine readable format;
- `myart.aux`, the auxiliary file; it is used by LATEX for internal "book keeping", including cross-referencing;
- `myart.log`, the log file; LATEX records the typesetting session in the log file, including the warnings and the errors.

Use a video driver to display the typeset article, `myart.dvi`, on the monitor, and a printer driver to print the typeset article, `myart.dvi` on a printer.

It should be emphasized that of the four programs used, only one (TEX) is the same for all computers and all implementations. If you use TEX in an "integrated environment", then all four programs appear as one.

1.11.1   *$\mathcal{AMS}$-LATEX revisited*

Now that you understand the structure of LATEX, we can again discuss $\mathcal{AMS}$-LATEX, a set of enhancements to LATEX by the $\mathcal{AMS}$. As outlined in the Introduction, the $\mathcal{AMS}$ enhancements to LATEX fall into three groups: the $\mathcal{AMS}$ math enhancements, the document classes, and the AMSFonts. Each consists of several packages.

An $\mathcal{AMS}$ document class automatically invokes the following $\mathcal{AMS}$ packages (see section 8.5 for a more detailed discussion and for the package interdependency diagram, Figure 8.3):

- amsmath, the main $\mathcal{AMS}$ math package;
- amsthm, proclamations with style and the `proof` environment;
- amsopn, operator names;
- amstext, the `\text` command;
- amsfonts, commands for math alphabets;
- amsbsy, bold symbol commands.

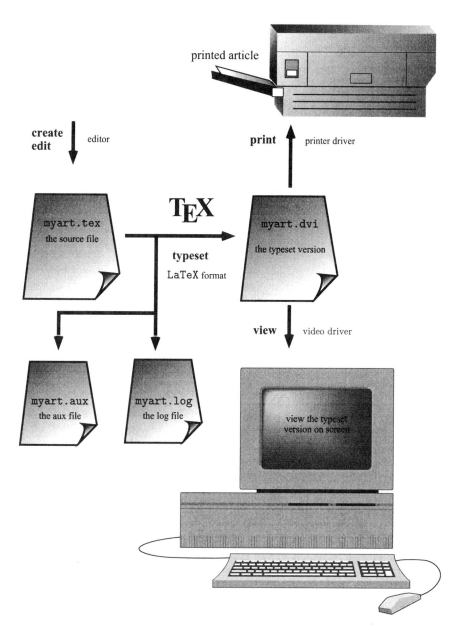

Figure 1.3: Using LᴬTᴇX

They do not automatically input the amssymb package, which provides the math symbol names. You can additionally input this and other $\mathcal{A}_{\mathcal{M}}\mathcal{S}$-LaTeX or LaTeX packages as needed.

When we discuss a feature of LaTeX that requires a package, I point this out in the text. I do not always point out, however, the interdependencies of the document classes and of the packages. For instance, the \text command (section 2.9) is provided by the amstext package, which is loaded automatically by the amsmath package, which in turn is loaded automatically by each of the $\mathcal{A}_{\mathcal{M}}\mathcal{S}$-LaTeX document classes. These interdependencies are discussed in section 8.5.

1.11.2   Interactive LaTeX

As a rule, LaTeX typesets an article non-interactively. Occasionally, you may want to use LaTeX interactively, that is, give LaTeX an instruction and ask it to carry it out. If LaTeX can't carry out your instructions, it displays a prompt:

- The ** prompt means that LaTeX wants to know the name of a source file to typeset. Probably, you misspelled a name, or you are in the wrong directory.
- The ? prompt asks "What should I do about the error I found?" Press return to continue; most of the time LaTeX recovers from the error, and completes the typesetting. If LaTeX can't recover from the error at the ? prompt, press X to exit. Typing H instead may yield useful advice.
- The * signifies interactive mode: LaTeX is waiting for an instruction. To get to such a prompt, comment out the line

\end{document}

(by inserting % as the first character of the line) in the source file and typeset. Interactive instructions (such as \show—see section 9.1.6) may be given at the * prompt. Typing

\end{document}

at the * prompt exits LaTeX.

1.11.3   Files

A number of files are created when a document called, say, myart.tex is typeset. When the typesetting takes place, a number of messages appear on the monitor. These are stored in the *log file*, myart.log. The typeset document is written in the myart.dvi file. LaTeX also writes one or more auxiliary files, as necessary. The most important one is myart.aux, the aux file (see section E.2.4).

1.11.4 Versions

All components of LaTeX interact. Since all of them have many versions, make sure they are up-to-date and compatible. While writing this book, I used LaTeX 2_ε (LaTeX version 2e), issued on December 1, 1994. You can check the version numbers and dates by reading the first few lines of the files in an editor or by checking the dates shown on the file list discussed below.

LaTeX is updated every six months; in-between updates, the `ltpatch.ltx` document is posted periodically on the CTAN (see Appendix G). Get this file and place it in your TeX input directory. When you rebuild your formats, `ltpatch.ltx` will patch LaTeX.

When you typeset a LaTeX document, LaTeX introduces itself in the log file with a line such as

```
LaTeX2e <1994/12/01> patch level 3
```

giving you the release date and patch level. If you use a new feature of LaTeX that was introduced recently, place in the preamble of your document the command

```
\NeedsTeXFormat{LaTeX2e}[1994/12/01]
```

where the date is the release date of the version you must use.

As of this writing, $\mathcal{AMS}$-LaTeX is at version 1.2 and the AMSFonts font set is at version 2.2. See Appendix G on how to get updated versions of $\mathcal{AMS}$-LaTeX and the AMSFonts.

BibTeX is at version 0.99 (version 1.0 is expected soon). In this book, I use the `amsplain.bst` bibliographic style file (version 1.2a).

If you include the `\listfiles` command in the preamble of your document, the log file will contain a detailed listing of all the files used in the typesetting of your document.

Here are a few lines from such a listing:

```
*File List*
book.cls        1994/12/09 v1.2x Standard LaTeX document class
leqno.clo       1994/12/09 v1.2x Standard LaTeX option
bk10.clo        1994/12/09 v1.2x Standard LaTeX file (size option)
amsmath.sty     1995/02/23 v1.2b AMS math features
Ueus57.fd       1994/10/17 v2.2d AMS font definitions
latexsym.sty    1994/09/25 v2.1f Standard LaTeX package
xspace.sty      1994/11/15 v1.03 Space after command names (DPC)
Ulasy.fd        1994/09/25 v2.1f LaTeX symbol font definitions
***********
```

1.12   What's next?

Having read thus far, you probably know enough about LaTeX to write your first article. The best way to learn LaTeX is by experimentation. Later, you may want to read Parts II–V.

If you look at the source files of the sample articles, your first impression may be how very verbose LaTeX is. In actual practice, LaTeX is fairly easy to type. There are two basic tools to make typing LaTeX more efficient.

Firstly, you should have a good editor. For instance, you should be able to train your editor so that a single keystroke produces the text:

```
\begin{theorem} \label{T:}


\end{theorem}
```

with the cursor in the position following ":" (where you type the label).

Secondly, customizing LaTeX will make repetitious structures such as

```
\begin{equation}
   \langle \dots, 0, \dots, \overset{i}{d}, \dots, 0,
   \dots \rangle \equiv \langle
   \dots, 0, \dots, \overset{i}{c}, \dots, 0, \dots
   \rangle \pmod{\Theta},
\end{equation}
```

which prints

$$(3.1) \qquad \langle \dots, 0, \dots, \overset{i}{d}, \dots, 0, \dots \rangle \equiv \langle \dots, 0, \dots, \overset{i}{c}, \dots, 0, \dots \rangle \pmod{\Theta},$$

(see page 369) become much shorter and (with practice) more readable. Utilizing the user-defined commands \con (for congruence), \vct (for vector), and \gQ (for Greek theta), in `sampart2.tex` (in the `ftp` directory and in Appendix C), this formula becomes

```
\begin{equation}
   \con \vct{i}{d}=\vct{i}{c}(\gQ),
\end{equation}
```

which is about as long as the typeset formula itself.

The topic of user-defined commands is taken up in Part IV.

Finally, custom formats (section 9.7) substantially speed up the typesetting of an average document.

s not that I'm afraid

happens. Woody Allen

Woody Allen

Literature is news that STAYS

Literature is news that

$$\left(x^2 \begin{bmatrix} i,i+1 \\ \dfrac{i+3}{3} \end{bmatrix} \right)$$

egin{quote}

It's not that I'm afraid

there when it happens.

$\sqrt{\mu}$

$(i) - 2$

\emph{Woody Allen}

x^2

$i, i+1$

Literature is news t

$i+3$

quote}

PART II

Text and math

Typing text

In Chapter 1, we discussed very briefly how to type text into a document. Now we take up this topic more fully.

This chapter starts with keyboarding (section 2.1) and the space rules (section 2.2). Section 2.3 takes up a very important topic: how to instruct LaTeX with commands and environments. A document may contain symbols that can't be typed; in section 2.4, we discuss how to type these in text. The symbol % plays a special role in the source document: it's used to comment out lines; see section 2.5.

The source file of a LaTeX document is typed in one font, one size, and one shape; section 2.6 discusses the methods for changing fonts, their shapes, and sizes. In section 2.7, you learn about lines, paragraphs, and pages. The judicious use of horizontal and vertical spacing is an important part of formatting of a document. This is discussed in section 2.8. In section 2.9, you learn how to typeset text in an imaginary "box" as a single "large" character.

The chapter concludes with footnotes (section 2.10) and how to split up the source document into several files (section 2.11).

2.1   The keyboard

Most of the keys on the computer's keyboard produce characters; others are function or modifier keys.

2.1.1   The basic keys

The basic keys are grouped as follows:

Letters The 52 letter keys:

    a   b   c ... z   A   B   C ... Z

Digits The ten digits:

    1   2 ... 9   0

Old style digits are available with the \oldstylenums command; in the next line the default digits are followed by the old style digits:

1234567890   1234567890

Punctuation marks There are nine:

    ,   ;   .   ?   !   :   '   '   –

The first six are the usual punctuation marks. The ' is the *left single quote* (also known as the grave accent), while ' doubles as the *right single quote* and *apostrophe* (see section 2.4.1). The – key is the *dash* or *hyphen* (see sections 2.4.2 and 2.4.8).

Parentheses There are four:

    (   )   [   ]

(and) are *parentheses*; [and] are called *brackets*.

Math symbols There are seven:

    /   *   +   =   −   <   >

The minus sign − is typed as - (hyphen) in math mode (see section 4.4.1). The last three may only be typed in math mode. There is also a version of colon (:) for math formulas (see section 4.11).

Space keys Pressing the spacebar (or the tab key) gives the *space character*; pressing the *return* (or *enter*) key gives the *end-of-line character*. (These keys produce *invisible characters* that normally do not appear on your monitor.) Different computer systems have different end-of-line characters; this may cause some problems when transferring files from one system to another. A good editor will translate end-of-line characters automatically or on demand.

Sometimes it's important to know whether a space is required. In such cases, I'll use the symbol ␣ to indicate a space, for instance, \in␣ut and \␣.

The tilde ~ signifies *nonbreakable space* or *tie* and is discussed in section 2.4.3.

2.1.2 Special keys

There are thirteen special keys:

$ % & ~ _ ^ \ { } @ " |

They are mostly used to give instructions to LaTeX; some are used in math mode (see Chapter 4), and some in BibTeX (see Chapter 10). See section 2.4.4 on how to print these characters in text. Only @ requires no special instructions; type @ to print @.

Tip When sending LaTeX source files via e-mail, you should insure that all the special characters are transmitted correctly.

To check whether this is the case, e-mail a document containing all the characters and their names. The file might look something like this:

```
uppercase           ABCDEFGHIJKLMNOPQRSTUVWXYZ
lowercase           abcdefghijklmnopqrstuvwxyz
digits              0123456789
exclamation    !    double quote    "    number          #
tilde          ~    at              @    period          .
asterisk       *    plus            +    comma           ,
minus          -    backslash       \    forward slash   /
colon          :    semicolon       ;    less than       <
equals         =    greater than    >    question mark   ?
left paren     (    right paren     )    circumflex      ^
left bracket   [    right bracket   ]    underscore      _
left brace     {    right brace     }    vertical bar    |
left single quote  ‘    percent     %    ampersand       &
right single quote ’    dollar      $
```

You'll find this file as email.tex in the ftp directory.

2.1.3 Prohibited keys

Other than those discussed in sections 2.1.1 and 2.1.2, every other key is prohibited! Specifically, do not use the computer's modifier keys to produce special characters. LaTeX will either reject or misunderstand them.

TeX 3.0 (and later) provides the possibility of using some modified keys, however. Standards are being developed to ensure the portability of such source files: in December of 1994, the LaTeX3 team released a beta version of an input encoding package, inputenc, by Alan Jeffrey.

Tip   If there is a prohibited character in your document, you may receive the error message:

```
! Text line contains an invalid character.
1.222 completely irreducible^^?
                             ^^?
```

Delete and retype the offending word or line until the error goes away.

2.2   *Words, sentences, and paragraphs*

Text consists of words, sentences, and paragraphs. In text, *words* are separated by one or more spaces. A group of words terminated by a period, exclamation point, or question mark make a *sentence*. A group of sentences terminated by one or more blank lines constitute a *paragraph*.

2.2.1   *The spacing rules*

Here are the most important LaTeX rules about spaces in text:

Rule 1 ■ Spacing in text
Two or more spaces in text are the same as one.

Rule 2 ■ Spacing in text
Spaces, tabs, and end-of-line characters are equivalent.

Rule 3 ■ Spacing in text
A blank line (that is, two end-of-line characters separated only by blanks and tabs) indicates the end of a paragraph. \par indicates the same.

Rule 4 ■ Spacing in text

Spaces at the beginning of a line are ignored, unless the previous line ends with %
(see section 2.5).

Rules 1 and 2 make cutting and pasting text less error-prone. In the source
file, you do not have to worry about the line length or the number of spaces separating words, as long as there is one space or end-of-line character separating any
two words. So

```
You    do not have to     worry
 about the number of    spaces
separating words, as long as there
is   at least one space or end-of-line character
separating  any two words.
```

produces the same sentence as

```
You  do not have to worry about the number of spaces separating
words, as long as there is at least one space or end-of-line
character separating any two words.
```

However,

```
 the number of    spaces
separating words,
as long
```

and

```
 the number of    spaces
separating words
, as long
```

produce different outputs:

the number of spaces separating words, as long
the number of spaces separating words , as long

Observe the space between "words" and the comma in the second line; that space
is produced by the end-of-line character according to Rule 2.

Note, however, the importance of the readability of the source file (see section 1.2.4 for an example). LaTeX may not care about the number of spaces or line
length, but the reader (you, your coauthor, or an editor) may.

Rule 3 contradicts Rules 1 and 2; consider it an exception. Sometimes (especially when defining commands and environments—see sections 9.1 and 9.2) it's more convenient to indicate the end of a paragraph with \par.

Tip   When you e-mail a document, the lines may be cut off at character position 72. If you encounter such difficulties, keep the line length at 72 or less.

2.2.2   *The period*

LaTeX uses the spacing rules in section 2.2.1 to decide where to put a space when typesetting words and paragraphs. For sentences, the rules are slightly more complicated, however. LaTeX places a certain size space between words—the *interword space*—and a somewhat larger space between sentences—the *intersentence space*. So LaTeX has to decide whether or not a period indicates the end of a sentence.

Rule 1 ■ Period
A period after a capital letter (for instance, A. or CAT.) signifies an abbreviation or an initial. Every other period signifies the end of a sentence.

This rule works most of the time. When it does not, inform LaTeX, as stated in the following two rules.

Rule 2 ■ Period
If an abbreviation does not end with a capital letter (for instance, "etc.") and it's not the last word in the sentence, then follow the period by an interword space (\␣).

Recall that \␣ provides an interword space. Example:

```
introduce the second variable, etc.\ as required.   Also,\\
introduce the second variable, etc. as required.   Also,
```

will print:

introduce the second variable, etc. as required. Also,
introduce the second variable, etc. as required. Also,

Notice that "etc." in the first line is followed by a regular interword space. The intersentence space following "etc." in the second line is longer.

Rule 3 ■ Period

If a capital letter followed by a period is the end of a sentence, precede the period with \@.

Example:

```
This follows from condition~H\@.  Therefore, we can proceed\\
This follows from condition~H.  Therefore, we can proceed
```

will print:

This follows from condition H. Therefore, we can proceed
This follows from condition H. Therefore, we can proceed

Notice that there is not enough space after "H." in the second line.

To make the intersentence space equal to the interword space, use the command

```
\frenchspacing
```

To switch back to using spaces of different sizes, give the command

```
\nonfrenchspacing
```

2.3 Instructing LaTeX

How do you instruct LaTeX to do something special for you, such as starting a new line, changing emphasis, or displaying the next theorem? This is accomplished with *commands* and *environments*.

2.3.1 Commands and environments

The \emph *command* instructs LaTeX to emphasize text; the \@ command instructs LaTeX to insert an intersentence space after a period (section 2.2.2).

The flushright *environment* instructs LaTeX to right justify the text between the two lines

```
\begin{flushright}
\end{flushright}
```

The document environment contains the body of the article, and the abstract environment contains the abstract (if the document class provides for it).

Rule 1 ■ Commands and environments

An environment starts with the command

\begin{*name*}

and ends with

\end{*name*}

Between these two lines is the *body* of the environment, affected by the definition of the environment.

Rule 2 ■ Commands and environments

A LaTeX command starts with the backslash symbol \ and is followed by the *command name*. The *name* of a command is either a *single non-alphabetic character* (other than a tab or end-of-line character) or a *string of letters* (one or more letters).

So # and ' are valid command names (the corresponding commands \# and \' are used in sections 2.4.4 and 2.4.6, respectively), as are input and date. However, input3, in#ut, and in␣ut are not valid names (3, #, and ␣ should not occur in a multicharacter command name).

LaTeX has a few commands (for instance, $—see section 4.1) that do not follow this naming scheme, that is, they are not of the form *name*. See also section 9.4 for special commands with special termination rules.

Rule 3 ■ Commands and environments

LaTeX finds the end of a command name as follows:

- If the first character of the name is not a letter, the name is terminated after the first character.
- If the first character of the name is a letter, the command is terminated by the first non-letter.

If the command name is a string of letters, and is terminated by a space, then LaTeX discards all spaces following the command name.

So while input3 is an incorrect name, \input3 is not an incorrect command; it's just the \input command followed by the character 3, which is part of the following text or of the argument.

LaTeX also allows command names to be modified with *; they are called "*-ed commands". A large number of commands have *-ed variants; for instance, \hspace* (see section 2.8.1).

Rule 4 ■ Commands and environments
Command and environment names are *case sensitive*: `\ShowLabels` is not the same as `\showlabels`.

Commands may have *arguments*, typed in braces immediately after the command; the argument(s) are used in processing the command. Accents provide very simple examples. For instance, `\'{o}` (printing ó) consists of the command `\'` and the argument o (see section 2.4.6); and in `\bibliography{article1}`, the command is `\bibliography` and the argument is `article1` (see section 10.2.2).

Some environments have arguments; for example, the `alignat` environment (see section 5.4.2) may be invoked by the lines:

`\begin{alignat}{2}`

and

`\end{alignat}`

A command or environment may have more than one argument. The `\frac` command (see section 4.4.1) has two ($\frac{1}{2}$ prints $\frac{1}{2}$); the user-defined command `\con` (see section 9.1.2) has three. Some commands (and a few environments) have one or more *optional arguments*, arguments that may or may not be present.

Rule 5 ■ Commands and environments
An optional argument is enclosed in brackets `[ ]`.

For example, the `\sqrt` command (section 4.4.3) has an optional argument for roots other than the square root: `\sqrt[3]{25}` prints $\sqrt[3]{25}$; the `\documentclass` command (section 6.2) has an argument (the name of a document class) and an optional argument (a list of options).

Tip   If you get an error using a command, check:

1. the spelling of the command, including the use of uppercase and lowercase characters;
2. that all required arguments are given;
3. whether the optional argument is in brackets, not braces;
4. whether the command is properly terminated.

Most errors in the use of commands come from the termination rule. Let me first illustrate this with the `\today` command, which produces today's date;

you have already seen this command in section 1.1.4 (see also section 2.4.7). The correct use is:

`\today\␣is␣the␣day`

or

`\today{}␣is␣the␣day`

which both print

> May 19, 1995 is the day

In the first case, \today was terminated by \␣ a command that produces a space. If there is no space after \today as in

`\todayis␣the␣day`

you get the error message:

```
! Undefined control sequence.
1.4 {todayis
                 the day}
```

LATEX thinks that \todayis is the command, and of course, does not recognize it. If you type any number of spaces after \today (two, for example):

`\today␣␣is␣the␣day%Incorrect!`

LATEX interprets the spaces as one, by the first space rule, and will use that one space to delimit \today from the text that follows it. So LATEX prints

> May 19, 1995is the day

See section 9.4 on how to avoid such errors.

2.3.2 Scope

A command issued inside a pair of braces { } has no effect beyond the right brace. You can have any number of *non-overlapping* pairs of braces:

`{ ... { ... { ... } ... } ... }`

The innermost pair containing a command is the *scope* of that command; the command has no effect outside its scope (see, however, section 2.3.3). Let me illustrate this with the \bfseries command that switches the font to boldface:

```
{some text \bfseries bold text} no more bold
```

typesets the phrase "bold text" in bold, but "no more bold" is not typeset in bold:

some text **bold text** no more bold

The commands \begin{*name*} and \end{*name*} bracketing an environment (including inline and displayed math environments—see section 4.1), also act as a special pair of braces.

Remember the following two obvious but very important rules about braces:

Rule 1 ■ Braces

Braces must be balanced: an opening brace has to be closed, and a closing brace must have a matching opening brace.

Rule 2 ■ Braces

Pairs of braces cannot overlap.

Violating the first brace rule generates warnings and error messages. If there is one more opening brace than closing brace, the article is typeset, but you get a warning:

```
(\end occurred inside a group at level 1)
```

For two more opening braces, you are warned that \end occurred inside a group at level 2, and so on. There is a tendency to disregard such a warning since

- The article is already typeset.
- The error may be difficult to find.

However, such errors may have strange consequences. At one point in the writing of this book, there were two extra opening braces in Chapter 2; as a result, the title of Chapter 7 was placed on a page by itself. So it's best not to disregard such warnings.

If there is one extra closing brace, you get an error message of the type:

```
! Too many }'s
```

If a special brace (say, \begin{*name*}) does not balance, you get an error message as discussed in Examples 2 and 3 of section 1.8.

Here are two simple examples of overlapping braces:

Example 1

```
{\bfseries some text
\begin{lemma}
   more text} final text
\end{lemma}
```

Example 2

```
{some \bfseries text, then math: $\sqrt{2} }, \sqrt{3}$
```

In Example 1, the scope of \bfseries overlaps the special braces \begin{lemma} and \end{lemma}, while in Example 2, the scope of \bfseries overlaps the special braces $ and $. Example 1 is easy to correct:

```
{\bfseries... }
\begin{lemma}
   {\bfseries... }
   ...
\end{lemma}
```

Example 2 may be corrected as follows:

```
{some \bfseries text, then math:} $\sqrt{2}, \sqrt{3}$
```

If the braces do overlap, and they are of the same kind, LaTeX will simply misunderstand the instructions: the closing brace of the first pair will be regarded as the closing brace of the second pair, an error not easy to detect. Real conflicts develop when using special braces. For instance, Example 1 gives the error message:

```
! Extra }, or forgotten \endgroup.
l.7 more text }
                final text
```

2.3.3   *Types of commands*

It may be useful at this point to mention that commands may be of various types.

Some commands have arguments, and some do not. It's more interesting that some commands affect change on their arguments, while some commands declare a change. For instance, \textbf{This is bold} typesets the phrase "This is bold" in bold, and has no effect on the text following the command. On the other hand, the command \bfseries declares that the following text be bold; in fact, this command has no argument. I call a command that affects change a *command declaration*. So \bfseries is a command declaration, while \textbf is not. As a rule, command declarations are commands without arguments.

Commands with arguments are *long* if the argument(s) can be more than a paragraph long; otherwise they are *short*. For example, \textbf is a short command, and so are all the top matter commands discussed in section 8.2.

Finally, as discussed in section 2.3.2, the effect of a command remains within its scope. This is true only of *local* commands. There are also *global* commands; for instance, the \setcounter command in section 9.3.1.

2.4 Symbols not on the keyboard

A typeset document may contain a number of symbols that can't be typed; some may even be available on the keyboard, but you are prohibited from using them (see section 2.1.3). In this section, we discuss the commands that typeset some of these symbols in text.

2.4.1 Quotes

To produce single and double quotes

'subdirectly irreducible' and "subdirectly irreducible"

type

```
`subdirectly irreducible' and ``subdirectly irreducible''
```

Here ' is the left single quote and ' is the right single quote. Note that the double quote is obtained by pressing the single quote key twice, and **not** by using the double quote key. If you need single and double quotes together, as in "She replied, 'No.'", separate them with \, (which provides a thin horizontal space):

```
``She replied, `No.'\,''
```

2.4.2 Dashes

Dashes come in three sizes. The shortest dash (called a *hyphen*) is used to connect words:

Mean-Value Theorem

This is typed with a single dash:

```
Mean-Value Theorem
```

A medium-sized dash (called an *en-dash*) is typed as -- and is used for number ranges; for instance, the phrase "see pages 23–45", typed as see pages~23--45. (Note: ~ is a nonbreakable space or tie—see section 2.4.3).

Type:	Print:	Type:	Print:	Type:	Print:
\#	#	\$	$	\%	%
\&	&	\~{}	~	_	_
\^{}	^	\{	{	\}	}

Table 2.1: Special characters

A long dash is a punctuation mark—called an *em-dash*—used to mark a change in thought or to add emphasis to a parenthetical clause, as in this sentence. The two em-dashes in the last sentence are typed as follows:

```
punctuation mark---called an \emph{em-dash}---used
```

In math mode, a single dash becomes a minus sign − (see section 4.4.1).

2.4.3   Ties or nonbreakable spaces

A *tie* or *nonbreakable space* (sometimes called a *blue space*) is an interword space that can't be broken across lines; for instance, when referencing P. Neukomm in an article, you do not want the first initial P. at the end of a line and Neukomm at the beginning of the next line. To ensure this does not happen, type P.~Neukomm.

The following examples show some typical uses:

```
Theorem~1
Donald~E. Knuth
assume that $f(x)$ is (a)~continuous, (b)~bounded
```

Of course, if you add too many ~ symbols; for instance,

```
Peter~G.~Neukomm%Incorrect!
```

LaTeX may send you a message that the line is too wide (see section 2.7.1).

In an $\mathcal{AMS}$ document class, the ~ absorbs spaces, so typing P.␣~␣Neukomm is just as good; this is very convenient when you have to add the ~ during editing.

2.4.4   Special characters

The characters corresponding to nine of the thirteen special keys (see section 2.1.2) are produced by typing a backslash \ and then the key, as shown in Table 2.1.

If for some reason you want to print a backslash, type

```
$\backslash$
```

The key | is never used in text; if you need to print the math symbol |, type $|$ (see section 4.6.1).

Note that * prints * (and in math, it prints ∗), and @ prints @.

Finally, the key " should never be used in text; see section 2.4.1 for the proper way to typeset double quotes.

Tip   Be careful when typing \{ to print { and \} to print }. Typing { instead of \{ causes unbalanced braces, in violation of the first brace rule of section 2.3.2. Similar troubles develop if you type } instead of \}.

See section 2.3.2 for some consequences. You may avoid some of these problems by introducing user-defined commands as in section 9.5.

You can also print special characters with the \symbol command:

\symbol{94} prints ˆ

\symbol{126} prints ˜

\texttt{\symbol{92}} prints \

The argument of the \symbol command is a number: the position of the symbol in the "layout" (encoding) of the font; for the Computer Modern typewriter style font, the layout is shown in Table 2.2.

You can obtain similar tables for any font using the fonttbl.tex file in your ftp directory. The first table in this file is shown in section 3.7 as a simple example of the tabular environment. In section F.1, fonttbl.tex is used to display the layout of the Times font.

See also the nfssfont.tex file (in the standard LaTeX distribution, see section 7.3) documented in Chapter 7 of *The LaTeX Companion* (sections 7.3.4, 7.5.5, and Figure 7.8).

2.4.5   Ligatures

Certain groups of characters, when typeset, are joined together; such compound characters are called *ligatures*. There are five ligatures (if you use the Computer Modern fonts): ff, fi, fl, ffi, and ffl, which LaTeX typesets automatically.

If you want to prevent a ligature, put the last character of the group in braces. Compare iff with iff, typed as iff and if{f}, respectively (as in Formula 4 in the formula gallery—see section 1.3). You can also break up the ligature with an italic correction (see section 2.6.5), iff typed as if\/f. Neither solution is satisfactory all the time, though they work for short words LaTeX would not hyphenate. The constructs \mbox{f}l or f\kern0pt l always work.

2.4.6   Accents and symbols in text

LaTeX provides 15 European accents. Type the command for the accent (\ and a character), followed by the letter (in braces) on which you want the accent placed;

	0	1	2	3	4	5	6	7	8	9
0	Γ	Δ	Θ	Λ	Ξ	Π	Σ	Υ	Φ	Ψ
10	Ω	↑	↓	'	ı	¿	ı	J	`	´
20	˘	ˇ	¯	˙	¸	ß	æ	œ	ø	Æ
30	Œ	Ø	␣	!	"	#	$	%	&	'
40	(	)	*	+	,	-	.	/	0	1
50	2	3	4	5	6	7	8	9	:	;
60	<	=	>	?	@	A	B	C	D	E
70	F	G	H	I	J	K	L	M	N	O
80	P	Q	R	S	T	U	V	W	X	Y
90	Z	[	\	]	^	_	`	a	b	c
100	d	e	f	g	h	i	j	k	l	m
110	n	o	p	q	r	s	t	u	v	w
120	x	y	z	{	|	}	~	¨		

Table 2.2: Font table for Computer Modern typewriter style font

Type:	Print:	Type:	Print:	Type:	Print:	Type:	Print:
\`{o}	ò	\'{o}	ó	\"{o}	ö	\H{o}	ő
\^{o}	ô	\~{o}	õ	\v{o}	ǒ	\u{o}	ŏ
\={o}	ō	\b{o}	o̲	\.{o}	ȯ	\d{o}	ọ
\c{o}	ǫ	\r{o}	å	\t{oo}	o͡o		

Table 2.3: European accents

see Table 2.3.

Examples: To get Grätzer György, type

```
Gr\"{a}tzer Gy\"{o}rgy
```

To place an accent on top of 'i' or 'j', you must use the *dotless* 'i' and 'j', obtained by \i and \j. Examples: \'{\i} prints í and \v{\j} prints ǰ.

Tables 2.4 and 2.5 list some other symbols and European characters available in LaTeX when typing text.

Note that the \textcircled command has an argument; it seems to work best with a single lowercase character, for instance, ⓐ or ⓐ.

2.4.7 *Logos and numbers*

\TeX prints TeX, \LaTeX prints LaTeX, and \LaTeXe prints LaTeX 2$_\varepsilon$ (the current version of LaTeX). Remember to type \TeX\␣ or \TeX{} if you need a space after TeX (similarly for \LaTeX and \LaTeXe). A better way to handle this problem is discussed in section 9.1.1. The amsmath package adds the \AmS command, which

Type:	Print:	Type:	Print:	Type:	Print
\dag	†	\ddag	‡	\textbullet	•
\P	¶	\copyright	©	\pounds	£
\S	§	\textvisiblespace	␣	\textcircled{a}	ⓐ
		\textperiodcentered	·		

Table 2.4: Extra text symbols

Type:	Print:	Type:	Print:	Type:	Print:	Type:	Print:	Type:	Print:
\aa	å	\AA	Å	\ae	æ	\AE	Æ	\o	ø
\O	Ø	\oe	œ	\OE	Œ	\l	ł	\L	Ł
\ss	ß	\SS	SS	?`	¿	!`	¡		

Table 2.5: European characters

prints $\mathcal{AMS}$.

LaTeX also stores some useful numbers:

- \time is the time of day in minutes since midnight.
- \day is the day of the month.
- \month is the month of the year.
- \year is the current year.

Display these numbers with the \the command. Example:

```
The year: \the\year; the month: \the\month; the day: \the\day
```

prints:

> The year: 1995; the month: 5; the day: 19

Of more interest is the \today command, which prints today's date in the form: May 15, 1995. You may want to use this as the argument of \date in the $\mathcal{AMS}$ document classes (see Chapter 8). In some standard LaTeX document classes, for instance, in article, this is not necessary; leaving out the \date command typesets today's date on the title page of the document (see section 1.7.1).

Remember the termination rule (Rule 3 in section 2.3.1).

```
today's date in the form: \today (you may want
```

prints:

> today's date in the form: May 19, 1995(you may want

To get the desired effect, type \␣ after the date:

```
today's date in the form: \today\ (you may want
```

2.4.8   *Hyphenation*

LaTeX reads the source file one line at a time, until the end of the current paragraph; then it tries to balance the typeset lines (see section E.2.2). To achieve this, LaTeX hyphenates long words, using a built-in hyphenation algorithm and a database stored in the `hyphen.tex` file. You can help LaTeX do a better job, however.

Rule 1 ■ Hyphenation

If you find that LaTeX cannot properly hyphenate a word, put *optional hyphens* in the text. An optional hyphen is typed as \- ; this will allow LaTeX to hyphenate the word at this point (and only at this point) if the need arises.

Example: `data\-base`
Note that

- an optional hyphen prevents hyphenation at any other place in the word;
- placing an optional hyphen in a particular word does not affect any other occurrences of that word.

Rule 2 ■ Hyphenation

List the words that often need help in a command

```
\hyphenation{data-base as-so-ciate}
```

All occurrences of the listed words will be hyphenated if need be.

Please note that in the \hyphenation command the hyphens are designated by "-" and not by "\-", and the words are separated by spaces (not by commas). You must use optional hyphens for words with accented characters, e.g.,

```
Gr\"{a}t\-zer
```

Such words can't be included in a \hyphenation list (unless you use T1 encoding, which is not discussed in this book).

If you use an $\mathcal{A}_{\mathcal{M}}\mathcal{S}$ document class, a large number of such exceptional words are contained in a long \hyphenation list in the `cls` file; in particular, the word database is listed, so LaTeX will have no difficulty hyphenating it.

Rule 3 ■ Hyphenation

To *prevent* hyphenation, put the word in the argument of an \mbox command.

Example: type the word "database" as

`\mbox{database}`

if you do not want it hyphenated. Alternatively,

`\hyphenation{database}`

tells LaTeX not to hyphenate any occurrence of "database".

Tip   LaTeX does not hyphenate a hyphenated word except at the hyphen; nor does it hyphenate a word followed by an em-dash or en-dash (see section 2.4.2). For such words, LaTeX often needs help.

Sometimes a hyphen in a phrase should not be broken. For instance, the phrase "m-complete lattice" should not be broken after m; so type it as

`\mbox{$\mathfrak{m}$-com}\-plete lattice`

(See section 4.14.2 for \mathfrak.)

The amsmath package provides the \nobreakdash command:

`\nobreakdash-   \nobreakdash--   \nobreakdash---`

which prevents the break at the hyphen, en-dash, and em-dash, respectively. Example for en-dash:

`pages~24\nobreakdash--47`

Since LaTeX will not hyphenate a hyphenated word except at the hyphen, the command \nobreakdash- prevents the hyphenation of the whole word as though it were enclosed in an \mbox. The form

`\nobreakdash-\hspace{0pt}`

allows the normal hyphenation of the word that follows; for example,

`$\mathfrak{m}$\nobreakdash-\hspace{0pt}complete lattice`

allows `complete` to be hyphenated.

This coding of "m-complete lattice" is a natural candidate for a user-defined command (see section 9.1.1).

Tip   If you want to know how LaTeX would hyphenate a list of words, place it in the
            argument of the \showhyphens command.

For instance,

```
\showhyphens{summation reducible latticoid}
```

The result:

```
sum-ma-tion re-ducible lat-ti-coid
```

will be shown on the monitor and in the log file.

Tip   Some editors have a tendency to wrap lines in the source file by breaking them at
            a hyphen, introducing errors in the typeset document.

For instance,

```
It follows from Theorems~\ref{T:M} and \ref{T:Ap} that complete-
simple lattices are very large.
```

is typeset by LaTeX as follows

It follows from Theorems 2 and 5 that complete- simple lattices are very large.

As you can see, there is a space between the hyphen and the word "simple". Indeed, following the hyphen there is an end-of-line character, which was placed
there by the editor to break the line. By the second space rule (see section 2.2.1),
the end-of-line character was interpreted by LaTeX as a space. To correct the error,
make sure that there is no such line break, or comment out (see section 2.5) the
end-of-line character:

```
It follows from Theorems~\ref{T:M} and \ref{T:Ap} that complete-%
simple lattices are very large.
```

Better yet, rearrange the two lines:

```
It follows from Theorems~\ref{T:M} and \ref{T:Ap} that
complete-simple lattices are very large.
```

2.5 *Commenting out*

The % symbol makes LaTeX ignore the rest of the line. A typical use might be:

```
therefore, a reference to Theorem~1% check this!
```

which is a comment to yourself.

The % has many uses. For instance, a typical document class command (see section 7.1.2):

```
\documentclass[twocolumn,twoside,legalpaper]{article}
```

may be typed, with explanations, as

```
\documentclass[%
            twocolumn,%  option for two-column pages
            twoside,%    format for twosided printing
            legalpaper%  print on legal paper
            ]{article}
```

so that the undesired options may be commented out at a later time:

```
\documentclass[%
%           twocolumn,%  option for two-column pages
%           twoside,%    format for twosided printing
            legalpaper%  print on legal paper
            ]{article}
```

Notice that the first line is terminated with % to comment out the end-of-line character.

Tip Some command arguments do not allow spaces; if you want to split the line within an argument list, terminate the line with % as in the above example.

See also the example at the end of section 2.4.8.

It is useful to start an article with a comment identifying it, identifying the format, and the LaTeX version to be used:

```
% This is article.tex
% Typeset with LaTeX format
\NeedsTexFormat{LaTeX2e}[1994/12/01]
```

The third line specifies the December 1, 1994 (or later) release of LaTeX 2_ε; use such a declaration if you need some feature that was not available earlier.

| **Tip** | If the comment is too long, split it; otherwise, it may wrap to the next line. |

Of course, your editor may be smart enough not to do this, but what about the e-mail editor or the editor of a coworker? Long lines are a potential problem, so it's best to avoid them altogether.

Other uses of % include marking parts of the article for your own reference; for instance, commenting command definitions (see `lattice.sty` in section 9.5 and in your `ftp` directory). If something goes wrong inside a multiline math display (see Chapter 5), LaTeX does not tell you precisely where the error occurred. Try commenting out all but one line, until each line works separately.

Note that % does not comment out lines in a BIBTEX database document (see section 10.2.5).

| **Tip** | The 25% rule: if you want the % sign in text, make sure you type it as \%. Otherwise, % acts like a comment. There is no warning. |

Commenting out large blocks of text might be tedious with %. If you use the verbatim package, then try the `comment` environment:

```
\begin{comment}
   ...the commented out text...
\end{comment}
```

Rule 1 ■ comment environment
\end{comment} must be on a line by itself.

Rule 2 ■ comment environment
There can be no comment within a comment.

In other words,

```
\begin{comment}
   commented out text...
   \begin{comment}
      some more commented out text...
   \end{comment}
   and some more commented out text...
```

```
\end{comment}
```

is not allowed. There may be one of several error messages depending on the circumstances; for instance:

```
! Bad space factor (0).
<recently read> \@savsf
```

```
l.175 \end{comment}
```

The comment environment is very useful when working on a large document. Commenting out large parts you are not working on should speed up the typesetting. Remember to delete the comment environment before final typesetting.

Since the comment environment requires the verbatim package, include the line

```
\usepackage{verbatim}
```

in the preamble if you want to use it.

Tip The comment environment is also very useful in locating errors. Suppose you have unbalanced braces in the source file (see section 2.3.2). Working with a *copy* of the source file, comment out the first half at a safe point and typeset. If you still get the error message, the error is in the second half; delete the first half of the copy of the source file. If there is no error message, the error is in the first half; delete the other half of the copy of the source file. Proceed like this until you narrow down the error to a paragraph, which you can visually inspect.

2.6 Changing font characteristics

Although LaTeX chooses how to typeset characters, there are occasions when you want control over the shape or size of the font.

2.6.1 The basic font characteristics

You do not have to be a typesetting expert to recognize the following basic font attributes:

Shape Text normally is typeset upright, but for emphasis you may want to print *slanted* or *italicized* text. Occasionally, you may want to use the SMALL CAPS shape.

Monospaced and proportional Typewriters of old used *monospaced* fonts: the width of all the characters was the same. Most editors display text on a monitor using a monospaced font. LaTeX calls monospaced fonts *typewriter style*.

In this book, such a font is used to represent user input whereas normal text is typeset in a *proportional* font such that an "i" is narrow, an "m" is wide. Witness:

mmmmmm
iiiiii }(monospaced)

mmmmmm
iiiiii }(proportional)

Serif A *serif* is a small stroke used to finish off a main stroke of a letter, as at the top and bottom of the letter "M". Fonts without serifs are called *sans serif*. Here is a sample of the Computer Modern sans serif font: CM sans serif font. Sans serif fonts are used for titles and special emphasis. The standard Computer Modern serif font is the Computer Modern roman.

Series: weight and width The *series* is the combination of weight and width. The thickness of a stroke is the *weight*; how wide is the character is measured by the *width*.

light, *medium* (normal), and *bold* often describe weight.

narrow (condensed), *medium* (normal), and *extended* often describe width.

Computer Modern has bold fonts. Traditionally, when the user asks for bold CM fonts, *bold extended* (a somewhat wider version) is provided.

Size In most document classes, LaTeX prints text in 10 point size unless otherwise instructed. Larger sizes are used for titles, section titles, and so on. The abstract and footnotes are often set in 8 point size.

2.6.2   *The document font families*

In the document class, the style designer designates three document font families:

- the roman (upright and serifed) document font family,
- the sans serif document font family,
- the typewriter style document font family,

and picks one of these as the *document font family* or *normal family*. In all the examples in this book, the roman document font family is the document font family. When you use Computer Modern fonts in LaTeX, the three document font families are the Computer Modern roman, Computer Modern sans serif, and Computer Modern typewriter; the document font family is Computer Modern roman.

In this book, the roman document font family is Galliard; the sans serif document font family is Helvetica, and the typewriter style document font family is Computer Modern typewriter.

When typing a document, the document font family (normal family) is the default font. You can always switch back to it with:

Type	or	to switch to the
`\textnormal{...}`	`{\normalfont ...}`	document font family
`\textrm{...}`	`{\rmfamily ...}`	roman document font family
`\textsf{...}`	`{\sffamily ...}`	sans serif document font family
`\texttt{...}`	`{\ttfamily ...}`	typewriter style document font family

Table 2.6: Font family switching commands

`\textnormal{...}`

or with

`{\normalfont ...}`

Table 2.6 shows these two commands and three additional pairs of commands to help you switch among the three basic document font families.

2.6.3 *Command pairs*

As you can see, most commands that change font characteristics come in two forms.

- A command with an argument, like `\textrm{...}`, effects the change in its argument; these are short commands (they cannot contain two blank lines or the `\par` command, see section 2.3.3).
- A command declaration, like `\rmfamily` carries out the font change (within its scope, see section 2.3.2).

Use commands with arguments for small changes within a paragraph. They have two advantages:

- it's less likely you'll forget to change back to the normal font;
- you do not have to worry about italic corrections (see section 2.6.5).

Note: commands with arguments are required as modifiers by *MakeIndex* (see section 11.1).

For font changes involving more than one paragraph, use command declarations; these are also preferable in user-defined commands and environments (see Chapter 9).

2.6.4 *Shape commands*

There are five pairs of commands to change the font shape.

- `\textup{...}` or `{\upshape ...}` switch to the upright shape.
- `\textit{...}` or `{\itshape ...}` switch to the *italic shape*.
- `\textsl{...}` or `{\slshape ...}` switch to the *slanted shape*.

- \textsc{...} or {\scshape ...} switch to SMALL CAPITALS.
- \emph{...} or {\em ...} switch to *emphasis*.

The document class will specify how emphasis is to be typeset; normally, it's italic or slanted, but note that the emphasis is context-dependent. For instance,

\emph{Rubin space}

in the statement of a theorem is typeset as:

> *the space satisfies all three conditions, a so-called* Rubin space *that ...*

The emphasis changed the style of "Rubin space" from slanted to upright.

Tip   Be careful not to interchange the command pairs. For instance, if by mistake you type {\textit serif}, the result is: *s*erif. Only the "*s*" is italicized since \textit takes s as its argument.

2.6.5   *Italic correction*

The phrase:

> when using a *serif* font

may be typed as follows:

when using a {\itshape serif\/} font

The command \/ before the closing brace is called an *italic correction*. Look at the emphasized M in the next example: *M*M typed as {\itshape M}M; The first "*M*" is leaning over the second "M". To prevent this, add an italic correction. So "*M*M" should be typed as {\itshape M\/}M. Compare the above phrase with and without the italic correction:

> when using a *serif* font
> when using a *serif* font

The latter is not as pleasing to the eye.

Rule 1 ■ Italic correction

If the emphasized text is followed by a period or comma, the italic correction should be suppressed.

Example:

Do not forget. Do come

may be typed as

```
{\itshape Do not forget}.  Do come
```

Rule 2 ■ Italic correction

The shape commands *with arguments* do not require an italic correction; it is provided automatically.

So type the phrase "when using a *serif* font" the easy way:

```
when using a \textit{serif} font
```

Whenever possible, let LaTeX take care of the italic correction. However, if LaTeX is adding an italic correction where you feel it's not needed, you can override LaTeX with the \nocorr command. LaTeX will not add an italic correction before a period and a comma; these two punctuation marks are stored in the \nocorrlist command. By redefining this command, you can modify LaTeX's behavior.

Rule 3 ■ Italic correction

The italic correction is required with the commands: \itshape, \slshape, \em.

2.6.6 *Two-letter commands*

Users of LaTeX 2.09 and $\mathcal{AMS}$-LaTeX version 1.1 are accustomed to using the two-letter commands: \bf, \it, \rm, \sc, \sf, \sl, and \tt. These are not part of the new standard LaTeX. They are, however, still defined in LaTeX and $\mathcal{AMS}$-LaTeX document classes. The two-letter commands

1. switch to the document font family; and
2. then change to the appropriate shape.

There are a number of reasons not to use them:

- The two-letter commands are not part of the standard LATEX.
- The two-letter commands require manual italic correction.
- The two-letter commands do not behave the way the other LATEX font changing commands behave (see section 2.6.9): `\slshape \bfseries` is the same as
  `\bfseries \slshape` (slanted bold), but `\sl \bf` (bold) is not the same as `\bf`
  `\sl` (slanted).

2.6.7   Series

These attributes play a very limited role with the Computer Modern fonts. There
is only one important command pair:

`\textbf{...}` or `{\bfseries ...}`

for changes to bold (actually, bold extended). The commands:

`\textmd{...}` or `{\mdseries ...}`

which set both the weight and width to medium (normal) are seldom needed.

2.6.8   Size changes

Standard LATEX documents are typeset in 10 point size; if you need to typeset in
11 point or 12 point size use the 11pt or 12pt document class options (see section 7.1.2). The AMS-LATEX document classes also offer the 8pt and 9pt document class options (see section 8.4). The sizes of titles, subscripts, and superscripts
are automatically changed by the document class. If you must change the font size
(it's seldom necessary to do so), the following command declarations are provided:

`\tiny \scriptsize \footnotesize \small`

`\normalsize`

`\large \Large \LARGE \huge \Huge`

These are listed in increasing (nondecreasing) order (a document class may imple-
ment some adjacent commands in the same size). Some document classes have
additional size changing commands (see section 8.1.1).

For example, the size changing commands in the book document class with
the 10pt option (which is the default, see section 7.1.2) are implemented as follows:

\tiny	<small>sample text</small>
\scriptsize	sample text
\footnotesize	sample text
\small	sample text
\normalsize	sample text
\large	sample text
\Large	sample text
\LARGE	sample text
\huge	sample text
\Huge	sample text

2.6.9   Orthogonality

You are now familiar with the commands that change the font family, shape, series, and size. Each of these commands affects one and only one font attribute. For example, if you change the series, then the font family, shape, and size do not change; these commands act independently. In LaTeX terminology, the commands are *orthogonal*. From the user's point of view this has an important consequence: *the order of these commands does not matter*. So

\Large \itshape \bfseries

has the same effect as

\bfseries \itshape \Large

Note that this is not true of the two-letter commands of LaTeX 2.09, discussed in section 2.6.6.

Orthogonality also means that you can combine these font attributes any way you like: for instance, the commands

\sffamily \slshape \bfseries \Large

instruct LaTeX to change the font family to sans serif, the shape to slanted, the series to bold, and the size to \Large. If the corresponding font is not available, LaTeX will change it to a font that is available, and issue a warning. The font substitution algorithm (see section 7.6.3 of *The LaTeX Companion*) may not provide the font you really want, so the responsibility is yours to make sure that the necessary fonts are indeed available.

2.6.10   Boxed text

Boxed text is very emphatic: ⟨Do not touch!⟩ This is typed as follows:

```
\fbox{Do not touch!}
```

Boxed text cannot be broken, so if you want to frame more than one line of text, place it in the argument of a \parbox command or within a minipage environment (see section 2.9.2), which is then put in the argument of an \fbox command. For instance,

```
\fbox{\parbox[3in]{Boxed text cannot be broken,
so if you want to frame more than one line
of text, place it in the argument of a
$\backslash$\texttt{parbox}
command or within a
\texttt{minipage} environment}}
```

prints

> Boxed text cannot be broken, so if you want to frame more than one line of text, place it in the argument of a \parbox command or within a minipage environment

See section 4.17 for boxed formulas.

2.7 Lines, paragraphs, and pages

When typesetting a document, LaTeX breaks the text into lines, paragraphs, and pages. Sometimes you may want to influence how LaTeX does its work.

2.7.1 Lines

LaTeX typesets a document one paragraph at a time (see section E.2.2). It tries to split the paragraph into lines of equal width; if it fails to do that successfully, and a line is too wide, you get the overfull \hbox message. Here is a typical example:

```
Overfull \hbox (15.38948pt too wide) in paragraph at lines 11--16
[]\OT1/cmr/m/n/10 In sev-eral sec-tions of the course on ma-trix
the-ory, the strange term ``hamiltonian-
```

The log file records which lines are too wide. To see a warning in the typeset version as well, use the draft document class option:

```
\documentclass[draft]{article}
```

Lines that are too wide will be marked with a slug on the margin; a slug is a vertical bar of width \overfullrule.

Do not worry about such messages while writing the document. If you are preparing the final version and receive such a message, the first line of defense for an `overfull \hbox` is to see whether optional hyphens would help (see section 2.4.8). Read the warning message carefully to see which words LaTeX is unable to hyphenate properly. Barring that, a simple rephrasing of the paragraph often does the trick.

Recall that 72.27 points make an inch. So if the message indicates a 1.55812pt overflow, for instance, you may safely ignore it.

Tip If you do not want the 1.55812pt overflow reported whenever the document is typeset, enclose the offending paragraph (including the blank line indicating the end of the paragraph) with the lines

```
{\setlength{\hfuzz}{2pt}
```

and

```
}% end of \hfuzz
```

Choose the argument (2pt, in this case) to slightly exceed the reported error. This does not affect the typesetting, but the warning message—and the slug, with the `draft` option—is suppressed.

A somewhat more elegant way of doing this is to include the offending paragraph (including the blank line indicating the end of the paragraph) in a

```
\begin{setlength}{\hfuzz}{2pt}
\end{setlength}
```

environment.

You can force a linebreak in the middle of a paragraph with `\linebreak`. This command breaks the line at the point of insertion and stretches it; if LaTeX thinks that there was too little left on the line, you get an

```
Underfull \hbox (badness 4328) in paragraph at lines 8--12
```

message.

You can qualify `\linebreak` with an optional argument: 0 to 4. The higher the argument, the more it forces. The `\linebreak[4]` command is the same as `\linebreak`; `\linebreak[0]` allows the linebreak but does not force it.

The `\newline` command breaks the line but does not stretch it. The text after `\newline` starts at the beginning of the next line, without indentation.

The `\\` command is the same as `\newline`. It has two variants, which may be combined:

- `\\*` which prohibits a pagebreak following the line;

- \\[*length*], where *length* is the interline space you wish to specify, for instance, 12pt, .5in, or 1.2cm.

Note how the units are abbreviated. For example

```
Note how the units are abbreviated.\\[15pt]   For example
```

prints:

> Note how the units are abbreviated.
>
> For example

Since \\ can be modified by * or by [ ], LaTeX may get confused if after a \\ the next line starts with * or [. In such cases, type * as {*} or [as {[}. For instance, to get

> There are two sources of problems:
> [a] The next line starts with [.

type

```
There are two sources of problems:\\
{[}a] The next line starts with \texttt{[}.
```

If you fail to type {[}, you get the error message:

```
! Missing number, treated as zero.
<to be read again>
                        a
1.16 [a]
          The next line starts with \texttt{[}.
```

Rule ■ \\

The \\ command is the same as the \newline command *in text*, but not in environments or command arguments.

Finally, the *[*length*] form combines the two variants.

The \nolinebreak command plays the opposite role. The \nolinebreak[0] command is the same as \linebreak[0]; also, \nolinebreak[4] is the same as \nolinebreak.

The nolinebreak commands are seldom used, however, since ~ , \mbox (see section 2.4.3), and if available, \text (see section 2.4.8) accomplish the same thing most of the time.

Double spacing

It is convenient to proofread documents double spaced. Moreover, some journals demand that articles be submitted double spaced.

To typeset a document double spaced, include the command

\renewcommand{\baselinestretch}{2}

in the the preamble. Similarly, "line and a half" spacing is provided by

\renewcommand{\baselinestretch}{1.5}

Alternatively, get the setspace package, by George D. Greenwade, in the

tex-archive/macros/latex/contrib/other/misc

directory on the CTAN (see Appendix G). Invoke this package and in the preamble give the

\doublespacing

or the

\onehalfspacing

command declarations.

For more on this topic, see section 3.1.5 of *The LaTeX Companion*.

2.7.2 Paragraphs

Paragraphs are separated by a blank line or by the \par command. Error messages always indicate new paragraphs as \par. The \par form is very useful in user-defined commands and environments (see sections 9.1 and 9.2).

In some document classes, the first line of a paragraph is automatically indented. Indentation may be eliminated with \noindent or forced with \indent.

Sometimes—for instance, in a schedule, glossary, or index—you may want the first line of a paragraph not indented, and all the others indented by a specified amount. This is called a *hanging indent*, and is done by assigning the amount of indentation to the parameter \hangindent.

The following example from a LaTeX glossary illustrates this.

sentence is a group of words terminated by a period, exclamation point, or question mark.

paragraph is a group of sentences terminated by a blank line or by the \par command.

typed as

```
\setlength{\hangindent}{30pt}
\noindent
\textbf{sentence} is a group of words terminated by
 a period, exclamation point, or question mark.
```

```
\setlength{\hangindent}{30pt}
\noindent
\textbf{paragraph} is a group of sentences terminated by a
blank line or by the \verb|\par| command.
```

Notice that the \setlength command must be repeated for each paragraph.

Sometimes, you may change the value of the length command \hangafter, which specifies the number of lines not to be indented. The default value is 1. To change it to 2, use the command

```
\setlength{\hangafter}{2}
```

For more about the \setlength command, see section 9.3.2. Section 3.1.4 of *The LaTeX Companion* discusses the style parameters of a paragraph.

The preferred way of shaping a paragraph or a series of paragraphs is with custom lists; see section 9.6.

2.7.3 Pages

There are pagebreaking commands analogous to the line breaking commands discussed in section 2.7.1:

```
\newpage
\pagebreak
\pagebreak[0] to \pagebreak[4]
\nopagebreak
\nopagebreak[0] to \nopagebreak[4]
```

The \pagebreak[4] command is the same as \pagebreak.

The \nopagebreak[4] command is the same as \pagebreak.

The \nopagebreak[0] command is the same as \pagebreak[0].

Sometimes, when preparing the *final draft* of a document, you may want to add a line or two to a page to prevent it from breaking at an unsuitable line. You accomplish this with the \enlargethispage command. For instance,

```
\enlargethispage{\baselineskip}
```

will add a line to the page length; on the other hand,

```
\enlargethispage{-\baselineskip}
```

will take a line away. The

```
\enlargethispage{10000pt}
```

command will make the page very long, so you can break it wherever you wish with \pagebreak.

The *-ed version \enlargethispage* squeezes the page as much as possible.

There are special commands for allowing or forbidding pagebreaks in multi-line math displays (see section 5.8).

There are two more variants of the \newpage command. The

```
\clearpage
```

command does a \newpage and in addition prints all the figures and tables (see section 6.4.3) waiting to be typeset. The variant

```
\cleardoublepage
```

is used with the twoside document class option (see sections 7.1.2 and 8.4); it does a \clearpage and in addition makes the next printed page a right-hand (odd-numbered) page by inserting a blank page if necessary.

2.7.4 *Multicolumn printing*

LaTeX provides the twocolumn document class option for two-column printing (see sections 7.1.2 and 8.4). In addition, there is the \twocolumn command which starts a new page (by issuing a \clearpage) and then prints in double column format. An optional argument provides a two-column wide title. The \onecolumn command switches back to one-column format.

The multicol package by Frank Mittelbach provides a much more sophisticated mechanism, the multicols environment that can start in the middle of a page. The environment is invoked with:

```
\begin{multicols}{n}[title]
```

where *n* is the number of columns and *title* is an optional argument, a title. The `multicols` environment is customizable; see section 3.5.3 of *The LaTeX Companion*.

2.8   Spaces

The judicious use of horizontal and vertical spacing is an important part of the formatting of a document. Fortunately, most of the spacing decisions are done by the document class, but LaTeX has a large number of commands that allow the user to insert horizontal and vertical spacing. Use them sparingly, however.

2.8.1   Horizontal spaces

When typing text, there are three commands that are most often used to create (horizontal) white space (shown between the bars in the display below):

$$\backslash_{\sqcup}: \qquad\qquad \text{||}$$
$$\backslash\text{quad}: \qquad \text{| |}$$
$$\backslash\text{qquad}: \qquad \text{| |}$$

A \quad creates a 1 em and a \qquad a 2 em space (see section 2.8.3). The interword space created by _\sqcup can stretch and shrink. There are other commands that create smaller units of space. All the commands of section 4.11 can be used in ordinary text (see also Appendix B), but more appropriate are the \hspace and \phantom commands.

The \hspace command takes a length as a parameter. For example (recall that $|$ prints |):

Type:	Print:
`$|$\hspace{12pt}$|$`	| |
`$|$\hspace{.5in}$|$`	| |
`$|$\hspace{1.5cm}$|$`	| |

The command produces the same space as the space occupied by its typeset argument. Illustrations:

Type:	Print:
`$|$\need space$|$`	|need space|
`$|$\phantom{need space}$|$`	| |

and

```
alpha \phantom{beta} gamma \phantom{delta}\\
\phantom{alpha} beta \phantom{gamma} delta
```

prints:

alpha   gamma
        beta   delta

Horizontal space variant

At the beginning of each line (except at the beginning and end of a paragraph) LaTeX removes all spaces. This includes the removal of space produced by \hspace, \quad, and so on. The variant \hspace* creates a space that is not removed by LaTeX under any circumstances.

Example:

```
And text\\
\hspace{20pt}And text\\
\hspace*{20pt}And text
```

prints:

And text
And text
        And text

Use \hspace*, for instance, for customized indentation. To indent a paragraph, say by 24 points, give the command:

```
\noindent\hspace*{24pt}And text
```

So

```
And text\\
\noindent\hspace*{24pt}And text
```

prints:

And text
        And text

2.8.2 *Vertical spaces*

Vertical space may be required to make room for a picture or to add some interline space for emphasis. The latter, as you have seen in section 2.7.1, can be accomplished with the command \\[*length*]. Both goals are easily accomplished with the \vspace command which works just like \hspace (see section 2.8.1), except that it creates vertical space. Examples:

\vspace{12pt}, \vspace{.5in}, \vspace{1.5cm}.

Standard amounts of vertical space are provided by the three commands:

\smallskip, \medskip, \bigskip

These spaces depend on the style and the font size. In the style and font I am currently using, they represent a horizontal space of 3pt, 6pt, and 12pt, respectively (12pt is the distance from line to line in standard LaTeX documents with the default 10pt option).

Rule   ■   Vertical space commands
All vertical space commands add the vertical space *after* the typeset line in which the command appears.

To print

end of text.

New paragraph after vertical space

type

```
end of text.
```

```
\vspace{12pt}
```

```
New paragraph after vertical space
```

However, you will get

end of text. no new paragraph after vertical space and the issuing command

does not start a new paragraph

if you type:

```
end of text.
\vspace{12pt}
```

```
no new paragraph after vertical space and the issuing
command does not start a new paragraph
```

Vertical space variant

LaTeX removes vertical space at the beginning and end of each page. This includes the removal of space produced by \vspace. The variant \vspace* creates space that is not removed by LaTeX under any circumstances.

2.8.3 Relative spaces

The length of a space is usually given in *absolute units*: 12pt (points), .5cm (centimeters), 1.5in (inches). Sometimes, *relative units* are more appropriate, that is, units that are relative to the size of the letters in the current font. Such units are em and ex: 1em is approximately the width of an "M"; 1ex is approximately the height of an "x". (Historically, 1em was exactly the width of an "M"; this is no longer true. For instance, in the Computer Modern roman 10 point font, 1 em is 10.00002 points, while the width of "M" is 9.16669 points; and similarly for 1 ex.)

Examples:\hspace{12em} and \vspace{12ex}.

See also the \quad and \qquad commands in section 2.8.1.

2.8.4 Expanding spaces

The commands

\hfill, \dotfill, and \hrulefill

fill all available space in the line with spaces, dots, or a horizontal line, respectively. If there are two of these in the same line pushing against each other, the space is equally divided. These commands can be used to center text, to fill lines with dots in a table of contents, and so on. To obtain

2. Boxes 34

type

2. Boxes\dotfill 34

To print

ABC   and   ABC

type

```
ABC\hfill and\hfill ABC
```

To get

⌐
|
| ABC————————————————————————and————————————————————————ABC
└

type

```
ABC\hrulefill and\hrulefill ABC
```

For instance, in a centered environment—such as a \title (section 8.2.1) or a center environment (section 3.8)—you may use \hfill to flush a line right, as in:

⌐
|

                              This is the title
                                                                        First Draft

                                  Author

└

To achieve this, type

```
\begin{center}
   This is the title\\
   \hfill First Draft\\
   Author
\end{center}
```

2.9   Boxes

Sometimes it's useful to typeset text in an imaginary "box", and use this box as a single "large" character. A single-line box is made with the \mbox or the \makebox command, or with the \text command provided by the amstext or amsmath packages; a (multiline) box with a prescribed width is created with the \parbox command or with the minipage environment.

2.9.1   Line boxes

The \mbox command provides a "line box" that typesets its argument without line-breaks. The resulting box is handled by LaTeX as a "large" character. For instance,

```
\mbox{database}
```

typesets database and handles the eight characters as if they were one. This has a number of uses: it prevents LaTeX from hyphenating the word (see section 2.4.8) and permits the word to be used in math mode (see section 4.5).

The \text command, provided by the amsmath package, is an improved version of the \mbox command; its argument is typeset in the appropriate size, say, as subscript or superscript.

Line box—refinement

The \mbox command is the short form of the \makebox command. The full form of this command is

\makebox[*width*][*alignment*]{*text*}

where the arguments are

width the (optional) width of the box; if omitted, the box is as wide as necessary;

alignment one of c (the default), l, r, or s; l flushes the argument left; r right; the text is centered by default; s stretches the text the full length of the box if there is stretchable space in the argument;

text the text in the box.

A *width* argument can be specified in inches (in), centimeters (cm), points (pt), or in the relative measurement such as em (see sections 2.8.3 and 9.3.2).
Examples:

```
\makebox{Short title.}End\\
\makebox[2in][l]{Short title.}End\\
\makebox[2in]{Short title.}End\\
\makebox[2in][r]{Short title.}End\\
\makebox[2in][s]{Short title.}End
```

prints:

Short title.End
Short title.   End
              Short title.   End
                            Short title.End
Short   title.End

Four length commands can be used in the specification of *width*:

\height, \depth, \totalheight, and \width

which are the dimensions of the box that would be produced without the optional argument *width*. To be more specific, \height is the height above the baseline; \depth is the depth below the baseline; \totalheight is the sum of \height and \depth; and \width is the width of the box.

So to print "hello" in a box three times the width of the argument:

start\makebox[3\width]{hello}end

which prints:

start   hello   end

The \framebox command works exactly like \makebox, except that it draws a frame around the box:

\framebox[2in][l]{Short title}

prints:

Short title

Use this command to print the number 1 in a square box, as required in the title of [10]:

\framebox{\makebox[\totalheight]{1}}

which indeed prints:

1

Note that

\framebox[\totalheight]{1}

prints

which is not a square box. Indeed, \totalheight is the height of 1, which becomes the width of the box. The total height of the box, however, is the height of 1 to which you have to add twice the \fboxsep (the separation between the contents of the box and the frame, defined as 3 points) and twice the \fboxrule (the width of the line—rule, defined as 0.4 points).

2.9.2 *Paragraph boxes*

A paragraph box is like a paragraph: the text therein is wrapped into lines, but the width of the lines is set by the user.

The \parbox command typesets its second argument in a paragraph with a line width supplied as the first argument; the resulting box is handled by LaTeX as a "large" character. To print a column three inches wide:

> The \parbox command typesets its second argu-
> ment in a paragraph with a line width supplied as
> the first argument.

type

```
\parbox{3in}{The $\backslash$\texttt{parbox} command
typesets its second argument in a paragraph with a
line width supplied as the first argument.}
```

This is especially useful in a tabular environment; see "Refinements" in section 3.7 for multiline entries.

The width of the parbox can be specified in inches (in), centimeters (cm), points (pt), or in the relative measurement em (see section 2.8.3), to mention a few units. (See section 9.3.2 for a complete listing.)

Tip   The \parbox command requires two arguments. Dropping the first argument may give the error message

```
! Missing number, treated as zero.
<to be read again>
                    T
l.175
```

Dropping the second argument gives no error message but the result will most likely not be what you intended.

Paragraph box refinement

The "character" created by \parbox is placed on the line so that its vertical center is aligned with the center of the line. An optional first argument b or t forces the bottom or the top to be so aligned. For an example, see section 3.7.

The full syntax of \parbox is

```
\parbox[alignment][height][inner-alignment]{width}{text}
```

Similar to the \makebox command (section 2.9.1),

\height, \depth, \totalheight, and \width

may be used in the *height* argument.

The *inner-alignment* argument is the vertical equivalent of the *alignment* argument for \makebox, determining the position of *text* within the box; it may be any one of t, b, c, or s, denoting top, bottom, centered, or stretched alignment, respectively. When the *inner-alignment* argument is not specified, it defaults to *alignment*.

Paragraph box as an environment

The minipage environment is very similar to the \parbox command: it typesets the text of the environment using a line width supplied as the argument. It has an optional argument for bottom or top alignment, and the other \parbox refinements also apply. The difference is that the minipage environment can contain displayed text environments (see Chapter 3).

The minipage environment can also contain footnotes (see section 2.10) displayed within the minipage. See section 3.4.1 of *The LaTeX Companion* for complications that may arise from this.

2.9.3 Marginal comments

A variant of the paragraph box is used to make marginal comments; the command is \marginpar. For example:

\marginpar{Tricky computation!}

Do not use marginal comments in equations or multiline math environments.

Tip   Do not use too many marginal comments on any given page. LaTeX may misplace them.

If the document is printed two-sided, then the marginal comments are printed on the outside margin. The form

\marginpar[*left-comment*]{*right-comment*}

prints the required argument *right-comment* when the marginal comment prints on the right margin, and the optional argument *left-comment* when the marginal comment prints on the left margin.

The width of the paragraph box for marginal comments is stored in the length command

\marginparwidth

(see section 9.3.2 for length commands). If you want to change it, use

\setlength{\marginparwidth}{*new_width*}

for instance,

\setlength{\marginparwidth}{90pt}

This width is set by the document class. If you want to know the present setting, type

\the \marginparwidth

in the document and typeset it, or in interactive mode (see section 1.11.2) type

*\showthe \marginparwidth

See sections 3.4.2 and 4.1 of *The LaTeX Companion* for other style parameters pertaining to marginal notes.

2.9.4 *Solid boxes*

A solid filled box is made with the \rule command; the first argument is the width and the second is the height. For instance, to print

end of proof symbol: ■

type

end of proof symbol: \rule{1.6ex}{1.6ex}

In fact, you may notice that this symbol is usually slightly lowered:

end of proof symbol: ■

This is done with an optional first argument:

end of proof symbol: \rule[-.23ex]{1.6ex}{1.6ex}

Tip   If a command expects two arguments, and none or only one is supplied, LaTeX gives an error message.

For instance, \rule{1.6ex} will give the message:

```
! Paragraph ended before \@rule was complete.
```

or

```
! Missing number, treated as zero.
```

In the first error message, the reference to \@rule suggests that the problem is with the \rule command. So check the syntax of the \rule command, and you find that an argument is missing. The second error message is more suggestive, since there is, indeed, a missing number. Here is an example combining \rule with \makebox and \hrulefill:

```
1 inch:\quad\makebox[1in]{\rule{.4pt}{4pt}%
    \hrulefill\rule{.4pt}{4pt}}
```

which prints:

Solid boxes of zero width (and varying height) are called *struts*. Struts are invisible, but they force LaTeX to make room for them, changing the vertical alignment of lines. Struts are especially useful for fine-tuning formulas; see the end of section 3.7 and section 4.13 for examples.

Rule ■ 0 distance
Opt, 0in, 0cm, 0em all stand for zero width; 0 by itself is not acceptable.

For example, \rule{0}{1.6ex} gives the error message:

```
! Illegal unit of measure (pt inserted).
<to be read again>
                          h
1.251 \rule{0}{1.6ex}
```

2.9.5   *Fine-tuning boxes*

The command

```
\raisebox{displacement}{text}
```

typesets *text* in a box with a vertical *displacement*. If *displacement* is positive (resp. negative), the box is raised (resp. lowered).

The \raisebox command allows us to play games:

```
fine-\raisebox{.5ex}{tun}\raisebox{-.5ex}{ing}
```

prints: fine-^{tun}ing.

The \raisebox command has two optional arguments. For instance,

```
\raisebox{0ex}[1.5ex][.75ex]{text}
```

forces LaTeX to typeset *text* as if it extended 1.5ex above and .75ex below the line, resulting in a change in the interline space before and after the line. A simple version of this command, \smash, is discussed in section 4.13.

2.10 Footnotes

A footnote is written as the argument of a \footnote command. To illustrate the use of footnotes, we place one here,[1] typed as

```
\footnote{Footnotes are easy to place.}
```

If you want symbols to designate the footnotes, rather than numbers, give the command

```
\renewcommand{\thefootnote}{\ensuremath{\fnsymbol{footnote}}}
```

which provides up to nine symbols. See see section 9.1.1 for the \ensuremath command.

Section 3.4 of *The LaTeX Companion* shows how to customize footnotes.

In addition, there are the title page footnotes introduced by some document classes, for instance, by the \thanks and \date commands in the top matter of the amsart document class (see section 8.2 and the typeset title page footnotes on page 361).

2.10.1 Fragile commands

As a rule, LaTeX reads a paragraph of the source file, typesets it, and then goes on to the next paragraph (see section E.2). Certain parts of the source file, however, are typeset and stored for later use.

Examples include the title of an article, which is reused as a running head (section 8.2.1); titles of parts, sections, subsections, and other sectioning commands, which are used in the table of contents (sections 6.3.2 and 6.4.1); footnotes; table and figure captions (section 6.4.3), and index entries (Chapter 11).

These are *movable arguments*, and certain commands embedded in them must be protected from damage while being moved. LaTeX commands that need such protection are called *fragile*. The math mode \(and \) commands are fragile; $ is not.

---

[1] Footnotes are easy to place.

In a movable argument, fragile commands must be protected with a \protect command. So

```
The function \( f(x^{2}) \)
```

is not an appropriate section title, but

```
The function \protect \( f(x^{2}) \protect \)
```

is. Of course, so is

```
The function $f(x^{2})$
```

but to be on the safe side, protect every command that might cause problems in a movable argument. In section 6.3.2, you'll see an example of what happens if a fragile command is not protected.

2.11 *Splitting up the file*

Sometimes, it's convenient to write a LaTeX document in several pieces. There are two commands that combine separate files into one document.

2.11.1 *Input and include*

You can piece together a long document with the \input and \include commands. For example,

```
\input{subfile}
```

will insert the contents of the subfile.tex file as if it were typed at that place. You can also use the \include command for the same purpose; however, an \include-ed file always start on a new page.

For instance, if your document contains five chapters, chapter1.tex, and so on, and an appendix, appendix.tex, you bring the document together with the following commands:

```
\include{chapter1}
\include{chapter2}
\include{chapter3}
\include{chapter4}
\include{chapter5}
\include{appendix}
```

The \include command has some advantages over \input. The most important advantage is the use of the \includeonly command. If you are currently working on Chapter 4, put the

```
\includeonly{chapter4}
```

command in the preamble, and only Chapter 4 will be typeset; even so, the page numbers, section numbers, and cross-references will be correct. (That is not quite true. They are derived from the last typesetting of the other chapters—see section E.2.4.) You could also have

```
\includeonly{chapter4,chapter5}
```

Rule ■ File termination
It is recommended that an \endinput command terminate every file that is \input-ed or \include-ed.

If you terminate an \include-ed file with \end{document}, then the typeset document will terminate with a warning such as:

```
(\end occurred when \iftrue on line 6 was incomplete)
(\end occurred when \ifnum on line 6 was incomplete)
```

Remember that all included files start on a new page. If this does not suit you, merge the source files for final printing with cut and paste; as an alternative, change all \include commands to \input.

Tip   In the preamble, place the line

```
% \renewcommand{\include}{\input}
```

When you want to change all \include commands to \input, simply uncomment the above line (that is, remove the %).

For interactive inclusion of \include-ed files, use the askinclude package.

2.11.2   *Combining files*

If you want to e-mail your LaTeX document, oftentimes you have to send a number of additional command files, document class files, PostScript diagrams, and so on. LaTeX makes it easy to package all your files into one large file. Let's say that along with your main document, you want to send the lattice.sty command file and the myart.cls document class file.

To do this, place the following commands *before* the \documentclass command:

```
\begin{filecontents}{lattice.sty}
```

```
% Command file for lattice papers
   ....
   ....
\end{filecontents}

\begin{filecontents}{myart.cls}
% Document class myart.cls
% Use it with all Proceedings submissions
   ....
   ....
\end{filecontents}
```

When the document is typeset, the lines in each `filecontents` environment are written out in a file with the name given as the environment's argument, *provided that such a file does not already exist.* LaTeX informs you when it creates a file by writing an entry in the log file:

```
LaTeX Warning: Writing file 'lattice.sty'.
```

On the other hand, if the file `lattice.sty` already exists, you get the warning:

```
LaTeX Warning: File 'lattice.sty' already exists on the system.
               Not generating it from this source.
```

When, in the main document, LaTeX gets to the lines:

```
\documentclass{myart}
\usepackage{lattice}
```

the `myart.cls` document class and the lattice package are already available for input.

Text environments

There are three types of text environments in LaTeX:

1. displayed text environments: the text in the environment is displayed and usually separated from the text before and after by some vertical space;
2. text environments that create a "large symbol";
3. style and size environments.

We start with the general displayed text environments: lists (section 3.1), the `tabbing` environment (section 3.2), `quote` and others (section 3.3). Then we proceed to discuss some displayed text environments that are important in mathematical typesetting: proclamations (theorem-like structures), proclamations with style, and the `proof` environment (sections 3.4 and 3.5).

We discuss only one environment that produces a "large symbol": `tabular` (section 3.7). Finally, we discuss the style and size environments in section 3.8.

The examples from this chapter (along with the custom list examples from Chapter 9) are collected in the `textenv.tpl` file in the `ftp` directory (see page 4).

3.1 List environments

LaTeX provides three list environments ready for your use: `enumerate`, `itemize`, and `description`. See section 9.6 for custom lists.

In this section, the list environments will be formatted as they normally are by the book document class. Observe that in the rest of the book, they are formatted differently, as specified by the book designer.

3.1.1 Numbered lists: `enumerate`

A *numbered list* is created with the `enumerate` environment. Example:

This space has the following properties:

1. Grade 2 Cantor

2. Half-smooth Hausdorff

3. Metrizably smooth

Therefore, we can apply the Main Theorem ...

typed as

```
This space has the following properties:
\begin{enumerate}
    \item Grade 2 Cantor \label{Cantor}
    \item Half-smooth Hausdorff \label{Hausdorff}
    \item Metrizably smooth \label{smooth}
\end{enumerate}
Therefore, we can apply the Main Theorem \dots
```

Each item is introduced with an `\item` command. The numbers generated can be labeled and cross-referenced (see section 6.4.2). This construct can be used in theorems and definitions, and for listing conditions or conclusions. Note that the stylistic details are determined by the document class.

3.1.2 Bulleted lists: `itemize`

Example:

In this lecture, we set out to accomplish a variety of goals:

• To introduce the concept of smooth functions

- To show their usefulness in the differentiation of Howard-type functions

- To point out the efficacy of using smooth functions in Advanced Calculus courses

In addition to these mathematicians ...

typed as

```
In this lecture, we set out to accomplish a variety of goals:
\begin{itemize}
    \item To introduce the concept of smooth functions
    \item To show their usefulness in the differentiation
        of Howard-type functions
    \item To point out the efficacy of using smooth functions
        in Advanced Calculus courses
\end{itemize}
In addition to these mathematicians \dots
```

3.1.3 *Captioned lists:* description

Example:

In this introduction, we outline the history of this concept. The main contributors were

J. Perelman the first to introduce smooth functions.

T. Kovács who showed their usefulness in the differentiation of Howard-type functions.

A. P. Fein the main advocate of using smooth functions in advanced Calculus courses.

In addition to these mathematicians ...

typed as

```
In this introduction, we outline the history of this concept.
The main contributors were
\begin{description}
    \item[J. Perelman] the first to introduce smooth functions.
    \item[T. Kov\'acs] who showed their usefulness in the
        differentiation of Howard-type functions.
    \item[A. P. Fein] the main advocate of using smooth
```

```
        functions in advanced Calculus courses.
\end{description}
In addition to these mathematicians \dots
```

3.1.4   *Rule and combinations*

Rule ■ List environments
An \item must immediately follow
        \begin{enumerate}, \begin{itemize}, or \begin{description}.

If it does not, you get an error message. For instance,

```
\begin{description}
    This is wrong!
    \item[J. Perelman] the first to introduce smooth functions.
```

gives the error message:

```
! LaTeX Error: Something's wrong--perhaps a missing \item.

1.11    \item[J. Perelman]
                    the first to introduce smooth func...
```

Remember the rule for list environments, and check for text preceding the first \item.

Up to four list environments can be nested and mixed; for instance:

⌐

   1. First item of Level 1.

      (a) First item of Level 2.

         i. First item of Level 3.

            A. First item of Level 4.

            B. Second item of Level 4.

         ii. Second item of Level 3.

      (b) Second item of Level 2.

   2. Second item of Level 1.

Referencing the second item of Level 4: 1(a)iB

└

which is typed as

```
\begin{enumerate}
  \item First item of Level 1.
  \begin{enumerate}
    \item First item of Level 2.
    \begin{enumerate}
      \item First item of Level 3.
      \begin{enumerate}
        \item First item of Level 4.
        \item Second item of Level 4.\label{level4}
      \end{enumerate}
      \item Second item of Level 3.
    \end{enumerate}
    \item Second item of Level 2.
  \end{enumerate}
  \item Second item of Level 1.
\end{enumerate}
Referencing the second item of Level 4: \ref{level4}
```

Note that the label level4 collected all four counters (see section 6.4.2).

And here is a mixed version of this example:

1. First item of Level 1.

   - First item of Level 2.

      (a) First item of Level 3.
         - First item of Level 4.
         - Second item of Level 4.
      (b) Second item of Level 3.

   - Second item of Level 2.

2. Second item of Level 1.

Referencing the second item of Level 4: 1a

which is typed as

```
\begin{enumerate}
  \item First item of Level 1.
  \begin{itemize}
    \item First item of Level 2.
    \begin{enumerate}
      \item First item of Level 3.
```

```
        \begin{itemize}
          \item First item of Level 4.
          \item Second item of Level 4.\label{enums}
        \end{itemize}
        \item Second item of Level 3.
      \end{enumerate}
      \item Second item of Level 2.
    \end{itemize}
    \item Second item of Level 1.
  \end{enumerate}
  Referencing the second item of Level 4: \ref{enums}
```

Now the label enums collected the two enumerate counters (see section 6.4.2).

In all three types of list environments, \item may be followed by an optional argument, which will be displayed at the beginning of the item:

\item[*label*]

This form of the \item command is particularly useful in a description environment, whose items have no default label.

Tip   It may happen that the text following an \item starts with a [; this would cause LaTeX to think that \item has an optional argument. To make sure that this does not happen, type [as {[}. By the same token, if there is a] *inside* the optional argument, type it as {]}.

Section 3.2 of *The LaTeX Companion* explains how to customize the three list environments. For control over the style of the counter in the enumerate environment, use David Carlisle's enumerate package (see section 7.3.1).

3.2   *Tabbing environment*

Although of limited use for mathematical typesetting, the tabbing environment may be useful for typing algorithms, computer programs, etc. Whereas the width of a column is determined by LaTeX from the widest entry in a tabular environment (section 3.7), in the tabbing environment, the width of a column is under user control.

The \\ command is the line separator, tab stops are set by \= (and are remembered by LaTeX in the order they are given), and \> moves to the next tab position. If there already is a next tab position, \= resets the tab.

Lines of comments may be inserted with the \kill command (see the examples below) and with the % character. The difference is that a line with \kill can be used to set tab stops, while a commented out line can not be so used.

A simple example:

```
PrintTime
    Block[{timing},
          timing = Timing[expr];
          Print[ timing[[1]] ];
    ]
End[]
```

typed as

```
\begin{tabbing}
   Print\= Time\\
   \>Block\=[\{timing\},\\
   \>\>timing = Timing[expr];\\
   (careful with initalization)\kill
   \>\>Print[ timing[[1]] ];\\
   \>]\\
   End[\,]
\end{tabbing}
```

An alternative way to proceed is to use a line to set the tab stops, but then \kill the line so it does not print. We use the following example to illustrate this:

```
\begin{tabbing}
   \hspace*{.25in}\=\hspace{2ex}\=\hspace{2ex}\=\hspace{2ex}\kill
   \>   $k := 1$  \\
   \>   $l_k := 0$; $r_k := 1$  \\
   \>   \texttt{loop}  \\
   \>   \> $m_k := (l_k + r_k)/2$  \\
   \>   \> \texttt{if} $w < m_k$ \texttt{then}  \\
   \>   \>   \> $b_k := 0$; $r_k := m_k$  \\
   \>   \> \texttt{else if} $w > m_k$ \texttt{then}  \\
   \>   \>   \> $b_k := 1$; $l_k := m_k$  \\
   \>   \> \texttt{end if}  \\
   \>   \> $k := k + 1$  \\
   \>   \texttt{end loop}
\end{tabbing}
```

which prints

$k := 1$

$l_k := 0$; $r_k := 1$

```
loop
```
$$m_k := (l_k + r_k)/2$$
if $w < m_k$ then
$$b_k := 0;\ r_k := m_k$$
else if $w > m_k$ then
$$b_k := 1;\ l_k := m_k$$
end if
$$k := k + 1$$
```
end loop
```

Observe that there is no \\ command in a line containing the \kill command.

It's sometimes convenient to set the tabs in a \kill line with \hspace commands. Note that \> moves to the "next" tab stop, which, in fact, may mean to move back, and overprint.

There are about a dozen more commands peculiar to this environment; if you need to use tabbing often, please consult Chapter 5 of *The LaTeX Companion*.

3.3 *Miscellaneous displayed text environments*

There are four more displayed text environments: quote, quotation, verse, and verbatim. The quote environment is used for short (one paragraph) quotations:

> It's not that I'm afraid to die. I just don't want to be there when it happens. *Woody Allen*
>
> Literature is news that STAYS news. *Ezra Pound*

Typed as:

```
\begin{quote}
    It's not that I'm afraid to die.  I just don't want to be
    there when it happens.
    \emph{Woody Allen}

    Literature is news that STAYS news.
    \emph{Ezra Pound}
\end{quote}
```

Note that multiple quotes are separated by blank lines. In the quotation environment, the blank lines mark new paragraphs:

KATH: Can he be present at the birth of his child?

ED: It's all any reasonable child can expect if the dad is present at the conception.

Joe Orton

typed as

```
\begin{quotation}
   KATH: Can he be present at the birth of his child?

   ED: It's all any reasonable child can expect if the dad
   is present at the conception.
   \begin{flushright}
      \emph{Joe Orton}
   \end{flushright}
\end{quotation}
```

Here is an example of the verse environment:

I think that I shall never see
A poem as lovely as a tree.

Poems are made by fools like me,
But only God can make a tree.

Joyce Kilmer

typed as

```
\begin{verse}
   I think that I shall never see\\
   A poem as lovely as a tree.

   Poems are made by fools like me,\\
   But only God can make a tree.

   \begin{flushright}
      \emph{Joyce Kilmer}
   \end{flushright}
\end{verse}
```

Lines are separated by \\, and stanzas by blank lines. Long lines are typeset with the wrapped around part indented (hanging indent).

Finally, there is the verbatim text environment. You may need it if you write *about* TEX or some other computer program (most of the displayed source in this book was written in a verbatim environment) or if you need to include source code or user input in your writing. For instance, you may write to a journal about the article you are proofreading:

```
    Formula (2) in section 3 should be typed as follows:
\begin{equation}
    D^{\langle 2 \rangle} = \{\, \langle x_0, x_1 \rangle
        \mid x_0, x_1 \in D,\ x_0 = 0 \Rightarrow x_1 = 0 \,\}.
\end{equation}
Please make the corrections.
```

The problem is that if you just type

```
Formula (2) in section 3 should be typed as follows:
\begin{equation}
    D^{\langle 2 \rangle} = \{\, \langle x_0, x_1 \rangle
        \mid x_0, x_1 \in D,\ x_0 = 0 \Rightarrow x_1 = 0 \,\}.
\end{equation}
Please make the corrections.
```

TEX will typeset this:

Formula (2) in section 3 should be typed as follows:

$$(2) \qquad D^{\langle 2 \rangle} = \{\, \langle x_0, x_1 \rangle \mid x_0, x_1 \in D,\ x_0 = 0 \Rightarrow x_1 = 0 \,\}.$$

Please make the corrections.

If you want TEX to print the source exactly as typed, place it in the verbatim environment:

```
Formula (2) in section 3 should be typed as follows:
\begin{verbatim}
    \begin{equation}
        D^{\langle 2 \rangle} = \{\, \langle x_0, x_1 \rangle
            \mid x_0, x_1 \in D,\ x_0 = 0 \Rightarrow x_1 = 0 \,\}
    \end{equation}
\end{verbatim}
Please make the corrections.
```

Rule ■ verbatim text environment

- There can be no verbatim environment within a verbatim environment.
- There can be no verbatim environment within the argument of a command.

The violation of the first rule will result in environment delimiters that do not match. You'll get an error message of the type:

```
! \begin{document} ended by \end{verbatim}.
```

Tip There are two traps to avoid when using the verbatim environment.

- If the line \end{verbatim} starts with spaces, then a blank line is added to the typeset version.
- The characters following \end{verbatim} on the same line are dropped.

Type the last two lines of the above example as follows:

```
␣\end{verbatim}
Please make the corrections.
```

and look at the printed version:

> Formula (2) in section 3 should be typed as follows:
> ```
> \begin{equation}
>     D^{\langle 2 \rangle} = \{\, \langle x_0, x_1 \rangle
>     \mid x_0, x_1 \in D,\ x_0 = 0 \Rightarrow x_1 = 0 \,\}.
> \end{equation}
> ```
>
> Please make the corrections.

There is an unintended blank line before the last line.

To illustrate the second part of the Tip, type the last line of the above example as

```
\end{verbatim} Please make the corrections.
```

Then the text

```
Please make the corrections.
```

is dropped, and you receive a warning

```
LaTeX Warning: Characters dropped after
 '\end{verbatim}' on input line 17.
```

The verbatim environment also has an "inline" version called \verb. Here is an example:

```
Some European e-mail addresses also contain \texttt{\%};
recall that you have to type \verb+\%+ to get \texttt{\%}.
```

which prints:

Some European e-mail addresses also contain %; recall that you have to type \% to get %.

The character following the \verb command is a delimiter; in the example I used +. The argument starts with the character following the delimiter, and it is terminated by the next occurrence of the delimiter. So in the example, the argument is \%.

Choose the delimiter character carefully. For instance, if you want to typeset

```
$\sin(\pi/2 + \alpha)$
```

verbatim, and you type

```
\verb+$\sin(\pi/2 + \alpha)$+
```

then you get the error message:

```
! Missing $ inserted.
<inserted text>
                   $
1.5 \verb+$\sin(\pi/2 + \alpha
                               )$+
```

Indeed, the argument of \verb is $\sin(\pi/2 in this case, since the second + terminates the \verb command. Then LaTeX tries to print \alpha)$+ but can't because it's not in math mode. Use, for instance, ! in place of +:

```
\verb!$\sin(\pi/2 + \alpha)$!
```

Also, the entire \verb command must be on a single line. If it's not, as in

```
\verb!$\sin(\pi/2 +
\alpha)$!
```

you get the error message:

```
! LaTeX Error: \verb command ended by end of line..
```

```
1.6 \verb!$\sin(\pi/2 +
```

Rule ■ verb command

- The entire \verb command must be on a single line.
- There can be no space between the \verb and the delimiter.
- The \verb command cannot be contained in the argument of another command.
- The \verb command cannot be contained in an amsmath aligned math environment.

The \verb command has a *-ed version which prints the spaces as ␣. For example, \today␣␣the is typed as \verb*+\today the+.

An improved version of the verbatim environment (as well as the comment environment) is provided by the verbatim package; in fact, the rules discussed in this section are those of the verbatim package. The verbatim environment has some interesting variants; a number of them are discussed in section 3.3 of *The LATEX Companion*. For instance, they facilitate the typing of computer dialogues as in Appendix G of this book, and command syntax as illustrated on page 127.

3.4 *Proclamations (theorem-like structures)*

Major components of mathematical writing are theorems, lemmas, definitions, and so forth. In LATEX, these are typed in displayed text environments, called *proclamations* (some call them "theorem-like structures").

In the intrart.tex sample article (see pp. 39–40), there are two theorems, a definition, and a notation. These four environments have similar structure, only their names are different.

In the amsart sample article (see pp. 361–363), there are a number of different proclamations, in a variety of styles, with varying degrees of emphasis. Proclamations with style will be discussed in section 3.4.2.

You saw in section 1.5 that a proclamation is obtained in two steps.

Step 1. In the preamble, *define the proclamation* with a \newtheorem command. For instance, place in the preamble the lines

```
\newtheorem{theorem}{Theorem}
```

to define a theorem environment.

Step 2. *Invoke the proclamation* in the body of the article as an environment. Using the proclamation definition in Step 1, type

```
\begin{theorem}
    My first theorem
\end{theorem}
```

in the article to produce a theorem:

Theorem 1. My first theorem

In the proclamation definition,

```
\newtheorem{theorem}{Theorem}
```

the first argument, theorem, is the name of the environment that will invoke the theorem; the second argument, Theorem, is the name that will be printed. LaTeX will number the theorems automatically and print them with some vertical space before and after. It'll display **Theorem 1** in some nice form (depending on the document class) followed by the theorem itself, probably emphasized (again, depending on the document class).

When you type the environment, it may have an optional argument; for instance,

```
\begin{theorem}[The Fuchs-Schmidt Theorem]
    The statement of the theorem
\end{theorem}
```

will print:

Theorem 2. (The Fuchs-Schmidt Theorem) The statement of the theorem

LaTeX is very fussy about how proclamations are defined. For instance, in the introductory article intrart.tex (see section 1.5)

```
\newtheorem{theorem}{Theorem}
\newtheorem{definition}{Definition
\newtheorem{notation}{Notation}
```

if the closing brace is dropped from the end of the second line as shown, you get the error message:

```
Runaway argument?
{Definition \newtheorem {notation}{Notation}
! Paragraph ended before \@nthm was complete.
```

```
<to be read again>
                    \par
1.12
```

The line number is the end of the paragraph, so check all the \newtheorems to locate the source of the error.

Next, drop an argument. Change the second line of the same paragraph to:

```
\newtheorem{definition}
```

You get the error message:

```
! LaTeX Error: Missing \begin{document}.
```

```
1.11 \newtheorem{n
                  otation}{Notation}
```

The line

```
! LaTeX Error: Missing \begin{document}.
```

usually means that LaTeX got confused, and convinced itself that there is some text typed in the preamble, which should be moved past the line

```
\begin{document}
```

The mistake could be anywhere in the preamble. If you encounter such an error message, try to isolate the problem by commenting out parts of the preamble (see section 2.5).

Tip If a proclamation starts with a list environment, precede the list by \hfill.

Example:

```
\begin{definition} \label{D:prime} \hfill
   \begin{enumerate}
      \item $u$ is \emph{meet-irreducible} if
         $u = x \wedge y$ implies that
         $u = x$ or $u = x$.\label{mi1}
      \item $u$ is \emph{meet-irreducible} if
         $u = x \wedge y$ implies that
         $u = x$ or $u = x$.\label{mi2}
      \item $u$ is \emph{completely join-irreducible} if
         $u = \bigvee X$ implies that $u \in X$.
            \label{mi3}
   \end{enumerate}
\end{definition}
```

which prints

⌐

Definition 1.

1. *u is* meet-irreducible *if* $u = x \wedge y$ *implies that* $u = x$ *or* $u = y$.
2. *u is* join-irreducible *if* $u = x \vee y$ *implies that* $u = x$ *or* $u = y$.
3. *u is* completely join-irreducible *if* $u = \bigvee X$ *implies that* $u \in X$.

∟

Without the \hfill, "*1. u is* meet-irreducible ... " would print on the same line as "**Definition 1.**"

Consecutive numbering

Suppose you want the lemmas and propositions in your paper numbered consecutively. You accomplish this with the following two lines in the preamble:

```
\newtheorem{lemma}{Lemma}
\newtheorem{proposition}[lemma]{Proposition}
```

As you can see, the second \newtheorem command of the previous example has an optional argument, lemma, which is the environment name defined by the first \newtheorem command. This optional argument informs LaTeX to number propositions consecutively with lemmas. So in this example, lemmas and propositions will be consecutively numbered as **Lemma 1**, **Proposition 2**, **Proposition 3**, and so on.

The optional argument of a proclamation definition must be the environment name of a proclamation that *has already been defined*.

Numbering within a section

The \newtheorem command may have a different optional argument, as in the following example:

```
\newtheorem{lemma}{Lemma}[section]
```

which tells LaTeX to number the lemmas within sections. In section 1, there will be **Lemma 1.1**, **Lemma 1.2**, and so on; in section 2, **Lemma 2.1**, **Lemma 2.2**, and so on.

Instead of section, you may have any sectioning command appropriate for the document class: chapter, section, and subsection are most commonly used.

The last two features can be combined. For example:

```
\newtheorem{lemma}{Lemma}[section]
\newtheorem{proposition}[lemma]{Proposition}
```

sets up the lemma and proposition environments so that they are numbered consecutively within sections such as **Lemma 1.1**, **Proposition 1.2**, and **Proposition 2.1**, and so on.

3.4.1 The full syntax

The full form of \newtheorem is:

\newtheorem{*envname*}[*procCounter*]{*Name*}[*secCounter*]

where the two optional arguments are mutually exclusive, and

envname is the name of the environment to be used in the body of the document. For instance, you may use theorem for the *envname* of a theorem; so a theorem is typed inside a theorem environment. Of course, *envname* is just a label; you are free to choose any environment name, such as thm or george (as long as the name is not in use as the name of another command or environment). This also gives a name for the counter used by LaTeX to number these text environments.

procCounter is an optional argument; it equates the counter of the new proclamation with the counter of a previously defined proclamation. As a result, the two proclamations will be consecutively numbered.

Name is the text that is typeset when the proclamation is invoked. So if Theorem is given as *Name*, then in the document you get **Theorem 1**, **Theorem 2**, and so on.

secCounter is an optional argument that causes *Name* environments to be numbered within the appropriate sectioning units. So if theorem is the *envname* and section is the *secCounter*, then in section 1 there will be **Theorem 1.1**, **Theorem 1.2**, and so on; and in section 2, **Theorem 2.1**, **Theorem 2.2**, and so on. Proclamations may be numbered within subsections, sections, or chapters.

3.4.2 Proclamations with style

With the amsthm package (automatically loaded by all the $\mathcal{AMS}$-LaTeX document classes), you can choose one of three proclamation styles:

    plain, the most emphatic
    definition
    remark, the least emphatic

You also get a few extra options with amsthm. For example, the *-ed version of \newtheorem defines a proclamation that is not numbered.

The following lines show how the styles are chosen in the `sampart.tex` sample article (see page 364); see the typeset sample article (on pages 361–363) how the choosen styles effect the typeset proclamations.

```
\theoremstyle{plain}
\newtheorem{theorem}{Theorem}
\newtheorem{corollary}{Corollary}
\newtheorem*{main}{Main Theorem}
\newtheorem{lemma}{Lemma}
\newtheorem{proposition}{Proposition}


\theoremstyle{definition}
\newtheorem{definition}{Definition}


\theoremstyle{remark}
\newtheorem*{notation}{Notation}
```

The document class determines how the various styles are implemented; however, plain should be the most emphatic while remark is least emphatic.

Precede a \newtheorem command by the style setting \theoremstyle command. The default is the plain style. A \theoremstyle command stays in effect until another one is given; it has one argument: plain, definition, or remark.

Five examples

We present here five example sets of proclamation definitions. These examples are listed in the `amsart.tpl` file in the `ftp` directory (see page 4).

Example 1

```
\theoremstyle{plain}
\newtheorem{theorem}{Theorem}
\newtheorem{lemma}{Lemma}
\newtheorem{definition}{Definition}
```

The document may have theorems, lemmas, and definitions, each typeset in the most emphatic (plain) style. They are numbered separately, so that you may have **Definition 1**, **Definition 2**, **Theorem 1**, **Lemma 1**, **Lemma 2**, **Theorem 2**, and so on.

Example 2

```
\theoremstyle{plain}
\newtheorem{theorem}{Theorem}
```

```
\newtheorem{lemma}[theorem]{Lemma}
\newtheorem{definition}[theorem]{Definition}
\newtheorem{corollary}[theorem]{Corollary}
```

The document may have theorems, lemmas, definitions, and corollaries, type-set in the most emphatic (plain) style. They are all numbered consecutively, for example, **Definition 1**, **Definition 2**, **Theorem 3**, **Corollary 4**, **Lemma 5**, **Lemma 6**, **Theorem 7**, and so on.

Example 3

```
\theoremstyle{plain}
\newtheorem{theorem}{Theorem}
\newtheorem{proposition}{Proposition}[section]
\newtheorem{lemma}[proposition]{Lemma}
\newtheorem{definition}{Definition}

\theoremstyle{definition}
\newtheorem*{notation}{Notation}
```

The document may have theorems, propositions, lemmas, and definitions in the most emphatic (plain) style, and notations in the less emphatic definition style. Notations are not numbered. Propositions and lemmas are consecutively numbered *within sections*, so that you may have **Definition 1**, **Definition 2**, **Theorem 1**, **Lemma 1.1**, **Lemma 1.2**, **Proposition 1.3**, **Theorem 2**, **Lemma 2.1**, and so on.

Example 4

```
\theoremstyle{plain}
\newtheorem{theorem}{Theorem}
\newtheorem*{main}{Main Theorem}
\newtheorem{definition}{Definition}[section]
\newtheorem{lemma}[definition]{Lemma}

\theoremstyle{definition}
\newtheorem*{Rule}{Rule}
```

The document may have theorems, definitions, and lemmas in the most emphatic (plain) style, and unnumbered rules in the less emphatic (definition) style. Definitions and lemmas are numbered consecutively within sections. There is also an unnumbered Main Theorem. So, for example, you may have **Definition 1.1**, **Definition 1.2**, **Main Theorem**, **Rule**, **Lemma 1.3**, **Lemma 2.1**, **Theorem 1**, and so on.

Example 5

```
\theoremstyle{plain}
```

```
\newtheorem{theorem}{Theorem}
\newtheorem{corollary}{Corollary}
\newtheorem*{main}{Main Theorem}
\newtheorem{lemma}{Lemma}
\newtheorem{proposition}{Proposition}

\theoremstyle{definition}
\newtheorem{definition}{Definition}

\theoremstyle{remark}
\newtheorem*{notation}{Notation}
```

The document may have theorems, corollaries, lemmas, and propositions in the most emphatic (plain) style, and an unnumbered Main Theorem. It has definitions in the less emphatic (definition) style. All are separately numbered. For example, you may have **Definition 1**, **Definition 2**, **Main Theorem**, **Lemma 1**, **Proposition 1**, **Lemma 2**, **Theorem 1**, **Corollary 1**, and so on. Notations are unnumbered and typeset in the least emphatic (remark) style.

Number swapping

To number a proclamation on the left, for instance, **3.2 Theorem** (instead of the usual **Theorem 3.2**), put the \swapnumbers command before the \newtheorem command corresponding to the proclamation definition. So the proclamation definitions in the preamble should be in two groups: the regular ones should be listed first, followed by this command, then all the proclamations that swap numbers.

As an alternative to the amsthm package, you may wish to use Frank Mittelbach's theorem package (see section 7.3.1), from which amsthm evolved.

3.5   Proof environment

The amsthm package (automatically loaded by the $\mathcal{A}_{\mathcal{M}}\mathcal{S}$-LaTeX document classes) also defines a proof environment. Example:

Proof. This is the proof, delimited by the q.e.d. symbol.   □

typed as

```
\begin{proof}
   This is the proof, delimited by the q.e.d. symbol.
\end{proof}
```

The interline space separating the proof from the surrounding text is larger than normal, and the end of proof is marked with the symbol □ at the end of the line. There are a few examples of the proof environment in the sampart.tex sample article (pages 361–371).

If you do not wish a symbol at the end of a proof, for instance, if the proof ends with an equation, give the command:

```
\begin{proof}
    ...
    \renewcommand{\qedsymbol}{}
\end{proof}
```

To substitute for "*Proof*" another phrase, like "*Necessity*", as in

⌐

Necessity. This is the proof.   □

∟

use the proof environment with an optional argument:

```
\begin{proof}[Necessity]
    This is the proof.
\end{proof}
```

The optional argument may contain a reference, as in

```
\begin{proof}[Proof of Theorem~\ref{T:smooth}]
```

which may print

⌐

Proof of Theorem 5. This is the proof.   □

∟

3.6 *Some general rules for displayed text environments*

Blank lines in LaTeX play a special role, as they usually indicate the ends of paragraphs.

Since displayed text environments structure the printed display themselves, the rules about blank lines are somewhat relaxed. However, a trailing blank line signifies a new paragraph for the text following the environment.

Rule ■ Blank lines in displayed text environments

1. Blank lines are ignored immediately after \begin{*name*} and immediately before \end{*name*} except in a verbatim environment.
2. A blank line after \end{*name*} forces the following text to start a new paragraph.
3. As a rule, you should not have a blank line before \begin{*name*}.

The pagebreaking commands of section 2.7.3 apply to text environments, as does the \\ command of section 2.7.1 (including the use of the optional argument, as in \\[8pt]).

3.7 *Tabular environment*

A tabular environment creates a the table that is a "large symbol". Here is a sim-

ple table

Name	1	2	3
Peter	2.45	34.12	1.00
John	0.00	12.89	3.71
David	2.00	1.85	0.71

printed inline (which looks awful, but it

does make the point that the table is just a "large symbol"), typed as

```
\begin{tabular}{ | l | r | r | r | }
   \hline
   Name     & 1     & 2     & 3     \\ \hline
   Peter    &  2.45 & 34.12 & 1.00\\ \hline
   John     &  0.00 & 12.89 & 3.71\\ \hline
   David    &  2.00 & 1.85  & 0.71\\ \hline
\end{tabular}
```

with no blank line before or after the environment.

This table can be centered (with a center environment—see section 3.8). It can also be placed in a table environment (see section 6.4.3), and a caption may be added:

```
\begin{table}
   \begin{center}
      \begin{tabular}{ || l | r | r | r || }
         \hline
         Name     & 1     & 2     & 3     \\ \hline
         Peter    &  2.45 & 34.12 & 1.00\\ \hline
         John     &  0.00 & 12.89 & 3.71\\ \hline
         David    &  2.00 & 1.85  & 0.71\\ \hline
      \end{tabular}
      \caption{\protect\Times Tabular table}\label{Ta:first}
```

Name	1	2	3
Peter	2.45	34.12	1.00
John	0.00	12.89	3.71
David	2.00	1.85	0.71

Table 3.1: Tabular table

```
    \end{center}
\end{table}
```

This table is displayed as Table 3.1. It'll be listed in a list of tables (see section 6.4.3 and the front of this book) and the table number may be referenced by \ref{Ta:first}. Note that the label must be *between* the caption and the \end{table} command.

For another example, look at the two tables in the fonttbl.tex file in your ftp directory. The first is typed as

```
\begin{tabular}{r||l|l|l|l|l|l|l|l|l|l|} \hline \hline
  & 0 & 1 & 2 & 3 & 4 & 5 & 6 & 7 & 8 & 9\\\hline

0& \symbol{0} &\symbol{1}&\symbol{2}&\symbol{3}&
\symbol{4}&\symbol{5}&\symbol{6}&\symbol{7}&
\symbol{8}&\symbol{9}\\ \hline


. . . . . . . . .


120& \symbol{120} &\symbol{121}&\symbol{122}&\symbol{123}&
\symbol{124}&\symbol{125}&\symbol{126}&\symbol{127} & & \\ \hline
\end{tabular}
```

The second table is the same except that the numbers run from 128 to 255. The typeset tables are shown in sections 2.4.4 and F.1.

Rule ■ tabular environment

1. \begin{tabular} has an argument consisting of a character l, r, or c (meaning left, right, or center alignment) for each column, and the symbols |; each | indicates a vertical line in the table. Spaces in the argument are ignored (but use them for readability).

2. Columns are separated by & and rows are separated by \\.

3. & absorbs spaces on either side.

4. \hline signifies a horizontal line.

5. If you use a horizontal line to finish the table, \\ must separate the the last row of the table from \hline.
6. \begin{tabular} has an optional argument b or t for bottom or top vertical alignment. The default is center alignment.

Remember to put the optional argument b or t in brackets, as in

```
\begin{tabular}[b]{ || l | r | r | r || }
```

If you forget to place \\ at the end of the last row before \hline, you get the error message:

```
! Misplaced \noalign.
\hline ->\noalign
                  {\ifnum 0='}\fi \hrule \@height \arrayrulew...
1.9 ....00 & 1.85  & 0.71 \hline
```

More column formatting commands

The argument of the tabular environment may contain other column formatting commands.

An @-*expression*, for instance, @{.}, replaces the space LaTeX normally inserts between two columns. For example,

```
\begin{tabular}{r @{.} l}
   3&78\\
   4&261\\
   4
\end{tabular}
```

creates a table with two columns separated by a decimal point. In effect, you get a single, decimal-aligned column:

```
    3.78
    4.261
    4.
```

This is an illustration only, you should use David Carlisle's dcolumn package if you need a decimally aligned column (see section 7.3.1).

The width of a column depends on the entries in the column; if you want to change this, use the p column specifier:

```
p{width }
```

For instance, if you want the first column of Table 3.1 to be 1 inch wide, then type

```
\begin{tabular}{ || p{1in} | r | r | r || }\hline
   Name      &  1    & 2     & 3    \\ \hline
   Peter     &  2.45 & 34.12 & 1.00\\ \hline
   John      &  0.00 & 12.89 & 3.71\\ \hline
   David     &  2.00 & 1.85  & 0.71\\ \hline
\end{tabular}
```

which prints

Name	1	2	3
Peter	2.45	34.12	1.00
John	0.00	12.89	3.71
David	2.00	1.85	0.71

The items in the first column are placed flush left. To center them, precede *each* item with a \centering command (see section 3.8).

Refinements

\hline draws a horizontal line the whole width of the table; \cline{*a-b*} draws a horizontal line from column *a* to column *b*. For instance,

\cline{1-3}   or   \cline{4-4}

Another useful command is \multicolumn, which is a single entry for one or more columns; for example,

\multicolumn{3}{c ||}{\emph{absent}}

The first argument is the number of columns spanned by the entry, the second is the alignment (and optional vertical line designators | for this row only), and the third argument is the entry. An example is given in Table 3.2, typed as follows:

```
\begin{table}
   \begin{center}
      \begin{tabular}{ || l | r | r | r || } \hline
         Name    & 1    & 2     & 3\\ \hline
         Peter   & 2.45 & 34.12 & 1.00\\ \hline
         John    & \multicolumn{3}{c ||}{\emph{absent}}\\ \hline
         David   & 2.00 & 1.85  & 0.71\\ \hline
      \end{tabular}
      \caption{\protect\Times Floating table with
               $\backslash$\texttt{multicolumn}}
      \label{Ta:multicol}
```

Name	1	2	3
Peter	2.45	34.12	1.00
John	*absent*		
David	2.00	1.85	0.71

Table 3.2: Floating table with \multicolumn

```
  \end{center}
\end{table}
```

The next example, shown in Table 3.3, uses \multicolumn and \cline together:

```
\begin{table}
  \begin{center}
    \begin{tabular}{ || c  c   | c | r || } \hline
      Name  & Month & Week & Amount\\ \hline
      Peter & Jan.  & 1    & 1.00\\ \cline{3-4}
            &       & 2    & 12.78\\ \cline{3-4}
            &       & 3    & 0.71\\ \cline{3-4}
            &       & 4    & 15.00\\ \cline{2-4}
            & \multicolumn{2}{| l}{Total: } & 29.49\\  \hline
      John  & Jan.  & 1    & 12.01\\ \cline{3-4}
            &       & 2    & 3.10\\ \cline{3-4}
            &       & 3    & 10.10\\ \cline{3-4}
            &       & 4    & 0.00\\ \cline{2-4}
            & \multicolumn{2}{| l}{Total: } & 25.21\\  \hline
        \multicolumn{3}{|| l}{Grand Total:} & 54.70\\ \hline
    \end{tabular}
    \caption{\protect\Times Floating table with
             $\backslash$\texttt{multicolumn} and
             $\backslash$\texttt{cline}}
    \label{Ta:multicol+cline}
  \end{center}
\end{table}
```

The \parbox command (see section 2.9.2) can be used for multiline entries; recall that the first argument of \parbox is the width of the box. As an example, to replace "Grand Total" by "Grand Total for Peter and John", type the last line as:

```
\multicolumn{3}{|| l}{ \parbox[b]{10em}{Grand Total\\
for Peter and John:} } & 54.70\\ \hline
```

Name	Month	Week	Amount
Peter	Jan.	1	1.00
		2	12.78
		3	0.71
		4	15.00
	Total:		29.49
John	Jan.	1	12.01
		2	3.10
		3	10.10
		4	0.00
	Total:		25.21
Grand Total:			54.70

Table 3.3: Tabular table with \multicolumn and \cline

(note the use of the bottom alignment option—see section 2.9.2). The last row of the modified table prints

Grand Total
for Peter and John: 54.70

The spacing of "Grand Total" is not quite right. This can be adjusted with a "strut":

```
\parbox[b]{10em}{\rule{0ex}{2ex}Grand Total\\
            for Peter and John:}
```

where 2ex is the height of the strut (see section 2.9.4).

Finally, vertical spacing can be adjusted by redefining \arraystretch. For instance, in the table

	Area	**Students**
5th Grade:	63.4 m^2	22
6th Grade:	62.0 m^2	19
Overall:	62.6 m^2	20

typed as

```
\begin{center}
   \begin{tabular}{ | r | c | c | } \hline
      & \textbf{Area}  & \textbf{Students}\\ \hline
```

```
      \textbf{5th Grade}: & 63.4 \mbox{m$^{2}$} & 22\\ \hline
      \textbf{6th Grade}: & 62.0 \mbox{m$^{2}$} & 19 \\ \hline
      \textbf{Overall}: & 62.6 \mbox{m$^{2}$} & 20\\ \hline
   \end{tabular}
\end{center}
```

you may find that the rows are too crowded. This may be adjusted by adding the line

```
\renewcommand{\arraystretch}{1.25}
```

to the `tabular` environment; to limit its scope, add it after

```
\begin{center}
```

The adjusted table prints:

	Area	**Students**
5th Grade:	63.4 m^2	22
6th Grade:	62.0 m^2	19
Overall:	62.6 m^2	20

In some tables, horizontal and vertical lines do not always intersect as desired; fine control over intersections is provided by the hhline package (see section 7.3.1).

All of Chapter 5 of *The LATEX Companion* deals with tabular material. It discusses many extensions, including multipage tables, decimal point alignment, tables within tables, footnotes in tables, and so on.

3.8   *Style and size environments*

There are numerous text environments to set font characteristics. They have the same name as the corresponding command declarations:

```
      rmfamily  sffamily  ttfamily
   upshape  itshape  slshape  scshape  em
                 bfseries
```

For instance,

```
\begin{ttfamily} text \end{ttfamily}
```

prints *text* just like {\ttfamily *text* }. Remember to use the the command declaration names for the environment names; that is, textrm is not correct; use

rmfamily instead (see section 2.6.3). There are also text environments for changing the font size, from tiny to Huge (see section 2.6.8), and two additional font
size environments with the $\mathcal{A}_{\mathcal{M}}\mathcal{S}$-LaTeX document classes (see section 8.1.1).

Horizontal line adjustment of a paragraph is controlled by the environments
flushleft, flushright, and center. Within the flushright and center environments, it's customary to indicate linebreaks with the \\ command (see section 2.7.1). In the flushleft environment, you would normally let LaTeX wrap
the lines.

These above text environments can be used separately or in combination, as in

> The **simplest** text environments set the printing style and size.
> The commands and the environments have similar names.

typed as

```
\begin{flushright}
   The \begin{bfseries}simplest\end{bfseries}
   text environments set the
   printing style and size.\\
   The commands and the environments have similar names.

\end{flushright}
```

Note the blank line at the end delimiting the paragraph.

There are command declarations that correspond to these environments:

\centering centers

\raggedright left adjusts

\raggedleft right adjust

The effect of one of these commands is almost the same as that of the corresponding environment except that the environment places some vertical space before and
after the displayed paragraphs. For such a command declaration to affect the way a
paragraph is formatted, the scope must include the whole paragraph including the
blank line at the end of the paragraph.

CHAPTER

4

Typing math

TeX was designed for typesetting math; LaTeX further facilitates technical writing.

A math formula is typeset *inline*, that is, as part of the current line, or *displayed*, that is, on a separate line or lines with vertical space before and after the formula. A displayed formula is either typeset on a single line or on multiple lines. In this chapter we'll discuss formulas that are inline or displayed on a single line. Multiline math formulas will be discussed in Chapter 5.

This chapter starts with a discussion of the math environments (section 4.1), the spacing rules in math (section 4.2), and continues with the equation environment (section 4.3). The basic constructs of a formula: arithmetic, subscripts and superscripts, roots, binomial coefficients, integrals, and ellipses are discussed in detail in section 4.4. Another basic construct, text in math, is the subject of section 4.5.

The four important math topics: delimiters, operators, large operators, and math accents are dealt with in sections 4.6–4.9. In section 4.10, we discuss three types of "horizontal lines" above or below a formula that stretch: braces, bars, and arrows; there are also stretchable arrow math symbols.

LaTeX provides a large variety of math symbols. Section 4.11 classifies and describes them; section 4.12 discusses how to build new symbols from existing ones. Horizontal spacing commands in math are described in section 4.13. Math alphabets and symbols are discussed in section 4.14.

The amsmath package provides a variety of ways of numbering and tagging equations; these are described in section 4.15. Then section 4.16 discusses the generalized fraction command of the amsmath package. Finally, section 4.17 describes boxed formulas.

4.1 *Math environments*

A math formula is either *inline* such as $a \equiv b \pmod{\theta}$ and $\int_{-\infty}^{\infty} e^{-x^2} dx = \sqrt{\pi}$, or *displayed*:

$$a \equiv b \pmod{\theta}$$

and

$$\int_{-\infty}^{\infty} e^{-x^2} dx = \sqrt{\pi}$$

Notice how the appearance of the two formulas has changed going from inline to displayed form.

Inline and displayed math formulas are typeset using the *math environments* math and displaymath, respectively. Math formulas occur too often to be displayed with environments; therefore, LaTeX allows the use of the special braces \(and \) or $ as abbreviations for the math environment, and \[and \] for the displaymath environment. So the inline congruence may be typed as

```
$a \equiv b \pmod{\theta}$
```

and the displayed integral as

```
\[
  \int_{-\infty}^{\infty} e^{-x^{2}} \, dx = \sqrt{\pi}
\]
```

Using $ as a delimiter for a math environment is a bit of an anomaly, since the same character $ is both the opening and closing delimiter. This can easily cause trouble. Leave one $ out and LaTeX does not know whether an opening or a closing delimiter is missing. For instance,

```
Let $a be a real number, and let $f$ be a function.
```

would be interpreted by LaTeX as follows: "Let" is ordinary text. The string of characters

```
$a be a real number, and let $
```

indicates a math environment, whereas "f" is interpreted as ordinary text. Finally,

```
$ be a function.
```

is thought to be a `math` environment (opened by $) that is closed by the next $ in the paragraph. When you run out of dollar signs, you get the error message:

```
! Missing $ inserted.
```

and the line number shows the end of the paragraph. LaTeX will place a $ at the end of the paragraph and typeset it. For instance, our example will print:

Let *abearealnumber, andletfbeafunction.*

because a math environment ignores spaces. It is now obvious that a $ is missing after the first math letter *a*.

The same mistake using \(and \) is handled more elegantly by LaTeX.

```
Let \( a be a real number, and let \( f \) be a function
```

gives the error message:

```
! LaTeX Error: Bad math environment delimiter.

1.6 ...a real number, and let \(
                                    f \) be a function
```

LaTeX realizes that the first \(opened a `math` environment, so the second \(must be in error. In this case, it gives the correct line number.

In this book, I'll not use \(and \) to delimit inline math, however.

Rule ■ Math environments
No blank line is permitted in a `math` or `displaymath` environment.

If you violate this rule, LaTeX sends an error message:

```
! Missing $ inserted.
<inserted text>
                    $
...
1.7
```

where the line number indicates the first blank line.

Multiline math displays are also implemented as math environments; they are discussed in Chapter 5. Displayed text environments are the topic of Chapter 3. User-defined environments are discussed in section 9.2.

4.2 *The spacing rules*

In text, the most important spacing rule is that any number of spaces in the source file equals one space in the typeset document. The spacing rules for the math environments are even more straightforward:

Rule ■ Spacing in math

Spacing in math does not matter for LaTeX.

In other words, all spacing in math mode is done by LaTeX. For instance,

$a+b=c$

and

$a + b = c$

are both typeset as $a + b = c$. Of course, the space delimiting a command name from the text that follows can't, as a rule, be omitted. For instance, in

$a \quad b$

you can't drop the space between \quad and b. Remember also that if you start writing text in the argument of \mbox (or \text) in a formula, then the rules of spacing in text apply (see section 2.2.1).

By the above rule, spacing in math does not effect the printed version. However, keep in mind that you should

---

Tip Format the source file so that it's easy to read.

In the source file, it is good practice

- to place \[or \] on a line by itself;
- to leave spaces before and after binary operations and binary relations (including =);
- to indent environments so they stand out (indent by three spaces, for example).
- not to break formulas at the end of the line;

Develop your own distinctive style of writing math, and stick with it.

Tip The spacing in math and text after a comma is different; do not confuse the two.

Example 1. Type "$a, b \in B$" as
```
$a$, $b \in B$
```
and not as
```
$a, b \in B$
```

Example 2. Type "$x = a, b$, or c" as
```
$x = a$, $b$, or $c$
```
and not as
```
$x = a, b$, or $c$
```
Compare the last two typeset:

$x = a, b$, or c

$x = a, b$, or c

Example 3. Type "for $i = 1, 2, \dots, n$" as
"```for $i = 1$,~2, \dots,~n```"

4.3   *The equation environment*

An equation is a displayed math formula that is numbered by LaTeX.

Equations are typed in an `equation` environment. The `equation` environment and `displaymath` environment are exactly the same except that in the former the displayed formula is assigned a number:

$$(1) \qquad \int_{-\infty}^{\infty} e^{-x^2} \, dx = \sqrt{\pi}$$

This is typed as

```
\begin{equation} \label{E:int}
    \int_{-\infty}^{\infty} e^{-x^{2}} \, dx = \sqrt{\pi}
\end{equation}
```

By default, the LaTeX document classes place the equation numbers on the right, while the $\mathcal{AMS}$-LaTeX document classes place them on the left. The default choice of the document class can be overridden with a document class option (see sections 7.1.2 and 8.4). In this book the equation numbers will be shown on the left except for the typeset `intrart` sample article (and examples from it).

The `\label` command is optional in the equation environment. If there is a `\label` command, the number assigned to the equation can be referenced with the `\ref` command. Using the above example,

```
see (\ref{E:int})
```

prints: see (1). If you use the amsmath package, you can use the `\eqref` command, which places the parentheses automatically;

```
see \eqref{E:int}
```

also prints: see (1). In fact, it does more, it typesets the reference *upright*, even in italicized or slanted text. For more about cross-referencing, see section 6.4.2.

LaTeX numbers the equations consecutively. Exactly how it numbers the equations depends on the document class. As a rule, in articles, it numbers equations consecutively throughout, while in books, it starts from 1 within each chapter. If you use the amsmath package, you may choose to have the equations numbered within each section: (1.1), (1.2), ... in section 1; (2.1), (2.2), ... in section 2; etc. In the preamble (see section 6.2), include the command:

```
\numberwithin{equation}{section}
```

"Manual control" of numbering is discussed in section 9.3.1.

If you use the amsmath package, then the equation environment has a *-ed form, which suppresses the numbering; so

```
\begin{equation*}
    \int_{-\infty}^{\infty} e^{-x^{2}} \, dx = \sqrt{\pi}
\end{equation*}
```

prints the same as

```
\[
    \int_{-\infty}^{\infty} e^{-x^{2}} \, dx = \sqrt{\pi}
\]
```

There is however one difference; in the equation* environment you can still use \tag (see section 4.15).

Rule ■ Equation environment
No blank lines are permitted in an equation or equation* environment.

If you type:

```
\begin{equation} \label{E:int}
    \int_{-\infty}^{\infty} e^{-x^{2}} \, dx = \sqrt{\pi}

\end{equation}
```

LaTeX will give you the error message:

```
! Missing $ inserted.
<inserted text>
                $
1.8
```

On the other hand, if the amsmath package is present, the error message is

```
Runaway argument?
\def \\{\@amsmath@err {\Invalid@@ \\}\@eha } \label {E\ETC.
! Paragraph ended before \equation was complete.
<to be read again>
                        \par
1.8
```

4.4   Basic constructs

A formula is built up from various types of basic constructs. This section discusses
the following:

- Arithmetic operations
- Subscripts and superscripts
- Roots
- Binomial coefficients
- Integrals
- Ellipses

Other constructs are discussed in subsequent sections.

4.4.1   Arithmetic

The *arithmetic operations* are typed as expected. To get: $a+b$, $a-b$, $-a$, a/b, and
ab, type

```
$a + b$, $a - b$, $-a$, $a / b$, $a b$
```

respectively. If you prefer $\cdot$ or $\times$ for multiplication, use \cdot or \times, respectively; $a \cdot b$ is typed as $a \cdot b$ and $a \times b$ is typed as $a \times b$.

In displayed formulas, *fractions* are not normally typed with / but rather with
the \frac command. To get

$$\frac{1+2x}{x+y+xy}$$

type

```
\[
    \frac{1 + 2x}{x + y + xy}
\]
```

Fraction refinements

The amsmath package provides some refinements to the \frac command. You can
use displayed math style fractions with \dfrac and inline math style fractions with

\tfrac. Example:

$$\tfrac{3+a^2}{4+b} \qquad \dfrac{3 + a^2}{4 + b}$$

typed as

```
\[
    \tfrac{3 + a^{2}}{4 + b} \qquad \dfrac{3 + a^{2}}{4 + b}
\]
```

The \dfrac command is often used in matrices whose entries are normally typeset in displayed math style. See Formula 20 in the formula gallery (section 1.3) for an example, and also section 4.16.

4.4.2 *Subscripts and superscripts*

Subscripts are typed with _ and *superscripts* with ^. Remember to enclose in braces the expression to be subscripted or superscripted:

```
\[
    a_{1}, a_{i_{1}}, a^{2}, a^{b^{c}}, a^{i_{1}}, a_{i} + 1,
    a_{i + 1}
\]
```

prints:

$$a_1, a_{i_1}, a^2, a^{b^c}, a^{i_1}, a_i + 1, a_{i+1}$$

For a^{b^c}, type $a^{b^{c}}$, not a^{b}^{c}. If you type the latter, you get the error message:

```
! Double superscript.
1.6 $a^{b}^
              {c}$
```

Similarly, a_{b_c} is typed as $a_{b_{c}}$, not as a_{b}_{c}.

In many instances, the braces for the subscripts and superscripts can be omitted, but it's good practice for a beginner to type them anyway.

Tip   You may safely omit the braces for a subscript or superscript that is a single digit or letter, as in a_1, $(a + b)^x$, and so on, which print a_1, $(a + b)^x$. Be careful, however. If you have to edit a_1 to make it a_{12}, then the braces can be omitted no longer; you must type a_{12} to print a_{12} because a_12 prints $a_1 2$.

There is one symbol that is already superscripted in math mode, the *apostrophe* '. To get $f'(x)$ type `$f'(x)$`. However, f'^2 should be typed as `${f'}^{2}$`; typing it as `$f'^{2}$` gives the double superscript error if the amsmath package is used.

Sometimes, you may want a symbol to appear superscripted (or subscripted) by itself as in the phrase

use the symbol † to indicate the dualspace

typed as

```
use the symbol ${}^{\dagger}$ to indicate the dualspace
```

where {} is the *empty group*. It can be used to separate symbols, or as the base for subscripting and superscripting.

The \sb and \sp commands also typeset subscripts and superscripts, respectively, as in

```
$a \sb{1} - a \sp{x + y}$
```

which prints $a_1 - a^{x+y}$. These commands are seldom used, however, except in the alltt environment (see section 7.3) and in the *Mathematical Reviews*.

For multiline subscripts and superscripts, see section 4.8.2.

4.4.3 Roots

The \sqrt command produces a square root, for instance,

$\sqrt{5}$ prints: $\sqrt{5}$
$\sqrt{a + 2b + c^{2}}$ prints: $\sqrt{a + 2b + c^2}$

Here is a more interesting example:

$$\sqrt{1 + \sqrt{1 + \frac{1}{2}\sqrt{1 + \frac{1}{3}\sqrt{1 + \frac{1}{4}\sqrt{1 + \cdots}}}}}$$

typed as

```
\[
    \sqrt{1 + \sqrt{1 + \frac{1}{2}\sqrt{1 +
    \frac{1}{3}\sqrt{1 + \frac{1}{4}\sqrt{1 + \cdots}}}}}
\]
```

Roots other than the square root are done with an optional argument: to print $\sqrt[3]{5}$, for example, type `$\sqrt[3]{5}$`.

Root refinement

The placement of the optional parameter is not always very pleasing, witness $\sqrt[g]{5}$. So the amsmath package provides two additional commands:

```
\leftroot      \uproot
```

to move the root *left* or *right* (with a negative argument), and *up* or *down* (with a negative argument). You may find one of the following variants an improvement:

$\sqrt[g]{5}$ typed as   `$\sqrt[ \leftroot{2} \uproot{2} g ]{5}$`
$\sqrt[g]{5}$ typed as   `$\sqrt[ \uproot{2} g ]{5}$`

Experiment with the arguments of \leftroot and \uproot to find the best spacing for a particular root.

4.4.4   *Binomial coefficients*

In the amsmath package there is a command for binomial coefficients: \binom. Here are two examples inline: $\binom{a}{b+c}$ and $\binom{\frac{n^2-1}{2}}{n+1}$, and the same two examples displayed:

$$\binom{a}{b+c} \text{ and } \binom{\frac{n^2-1}{2}}{n+1}$$

The latter are typed as

```
\[
    \binom{a}{b + c}\text{ and }\binom{\frac{n^{2} - 1}{2}}{n + 1}
\]
```

You can use display style binomials inline with \dbinom, and inline style binomials in displayed math with \tbinom; for example: $\binom{a}{b+c}$ is typed as

```
$\dbinom{a}{b + c}$.
```

4.4.5   *Integrals*

You have already seen the formula $\int_{-\infty}^{\infty} e^{-x^2}\, dx = \sqrt{\pi}$ in both inline and displayed forms in the first section of this chapter. In the inline version, the lower limit is a subscript and the upper limit is a superscript. You can change this with the \limits command; to print $\int\limits_{-\infty}^{\infty} e^{-x^2}\, dx = \sqrt{\pi}$, type

```
$\int\limits_{-\infty}^{\infty} e^{-x^{2}} \, dx = \sqrt{\pi}$
```

and to print

$$\int_{-\infty}^{\infty} e^{-x^2}\, dx = \sqrt{\pi}$$

type

```
\[
    \int\nolimits_{-\infty}^{\infty} e^{-x^{2}} \, dx = \sqrt{\pi}
\]
```

See also the `intlimits` option of the amsmath package (section 8.4.1).
There are five variants of the basic integral:

$$\oint \quad \iint \quad \iiint \quad \iiiint \quad \idotsint$$

which print

$$\oint \quad \iint \quad \iiint \quad \iiiint \quad \int \cdots \int$$

Except for the first, all require the amsmath package.

For complicated bounds, use the `\substack` command or the subarray environment (see section 4.8.2).

The integral package

The $\mathcal{AMS}$ has released the experimental amsintx package, which provides enhancements of the amsmath package. It introduces new variants of the `\int` and `\sum` commands to better mark the boundaries of the expression under the integral or sum. It is a first attempt in LaTeX to introduce the idea of a markup to math formulas of this type.

4.4.6 Ellipses

The `\ldots` command produces the "low" *ellipsis* as in

$$F(x_1, x_2, \dots, x_n)$$

typed as

```
\[
    F(x_{1}, x_{2}, \ldots , x_{n})
\]
```

and `\cdots` provides the "centered" *ellipsis* in $x_1 + x_2 + \cdots + x_n$, typed as

```
$x_{1} + x_{2} + \cdots + x_{n}$
```

If you use the amsmath package, the `\dots` command will do what you want most of the time; amsmath picks out the symbol following `\dots` to decide whether to use low or centered ellipsis. If the decision reached by amsmath is not appropriate, you can force low dots with `\ldots` and centered dots with `\cdots`. For instance, amsmath typesets

```
\[
   F(x_{1}, x_{2}, \dots , x_{n})\quad
   x_{1} + x_{2} + \dots + x_{n}\quad
   \alpha(x_{1} + x_{2} + \dots)
\]
```

as

$$F(x_1, x_2, \ldots, x_n) \quad x_1 + x_2 + \cdots + x_n \quad \alpha(x_1 + x_2 + \ldots)$$

The first two are correct; to get the desired ellipsis for the third, type

```
\[
   \alpha(x_{1} + x_{2} + \cdots)
\]
```

which prints

$$\alpha(x_1 + x_2 + \cdots)$$

You can exercise even more control using the amsmath package with the following commands:

`\dotsc`, for dots with a comma

`\dotsb`, dots with a binary operation or relation

`\dotsm`, dots with multiplication

`\dotsi`, dots with integrals

These not only force the dots low or center, but also adjust the spacing.

See section 5.6.1 for an example of *vertical dots* with the `\vdots` command and *diagonal dots* with the `\ddots` commands.

4.5 Text in math

LaTeX allows you to type text in formulas with the `\mbox` command. For instance,

$$A = \{\, x \mid x \in X_i \text{ for some } i \in I \,\}$$

is typed as

```
\[
   A = \{\, x \mid x \in X_{i} \mbox{ for some } i \in I \,\}
\]
```

Note that you have to leave space before "for" and after "some" inside the argument of \mbox. The argument of \mbox is always typeset in one line.

Sometimes it's convenient to go into math mode within the argument of \mbox:

$$A = \{ \, x \mid \text{for } x \text{ large} \, \}$$

which is typed as

```
\[
   A = \{\, x \mid \mbox{for $x$ large} \,\}
\]
```

The preferred way to enter text in formulas is with the \text command, provided by the amstext package. This package is automatically loaded by the amsmath package and by the $\mathcal{AMS}$-LaTeX document classes.

The \text command is similar to the \mbox command except that \text correctly sizes the text. The formula

$$a_{\text{left}} + 2 = a_{\text{right}}$$

is typed as

```
\[
   a_{\text{left}} + 2 = a_{\text{right}}
\]
```

Note that \text typesets the argument *in the size and shape* of the surrounding text. If you want the text in a formula typeset in the document font family (see section 2.6.2) independent of the surrounding text, use

```
\textnormal{ ... }
```

All the text font commands with arguments (see section 2.6.4) can be used in math formulas. For instance, \textbf will use the size and shape of the surrounding text to typeset its argument in bold (extended). The \textbf command (and the others) is like the \mbox command provided the amstext package is not loaded, that is, it does not change the size of its argument in a math formula. However, if the amstext package is loaded, \textbf changes size in a math formula like \text.

4.6   *Delimiters*

Delimiters are parenthesis-like symbols that bracket a formula, for example:

$$\left(\frac{1}{2} \right)^{\alpha}$$

and $3(a + b)$. Delimiters in LaTeX come in two varieties: fixed size and variable size. The latter stretch to enclose the formula.

Name:	Type:	Print:	Name:	Type:	Print:
Left paren.	(	(	Right paren.	)	)
Left bracket	[	[	Right bracket	]	]
Left brace	\{	{	Right brace	\}	}
Reverse slash	\	\	Forward slash	/	/
Left angle br.	\langle	⟨	Right angle br.	\rangle	⟩
Vertical line	|	|	Double vert. line	\\|	‖
Left floor br.	\lfloor	⌊	Right floor br.	\rfloor	⌋
Left ceiling br.	\lceil	⌈	Right ceiling br.	\rceil	⌉
Upper left corner	\ulcorner	⌜	Upper right corner	\urcorner	⌝
Lower left corner	\llcorner	⌞	Lower right corner	\lrcorner	⌟

Table 4.1: Standard delimiters

Name:	Type:	Print:
Upward arrow	\uparrow	↑
Double upward arrow	\Uparrow	⇑
Downward arrow	\downarrow	↓
Double downward arrow	\Downarrow	⇓
Up-and-down arrow	\updownarrow	↕
Double up-and-down arrow	\Updownarrow	⇕

Table 4.2: Arrow delimiters

4.6.1 Delimiter tables

The standard delimiters are shown in Tables 4.1 and 4.2. The "corners" in Table 4.1 require the amsmath package, however. Two synonyms are not shown: \vert is the same as |; the \Vert command is the same as the \| command (printing ‖).

The delimiters |, \|, and all the arrows of Table 4.2 are special: the same symbol represents both the left delimiter and the right delimiter. If this causes problems, use \left and \right to tell LaTeX whether the delimiter is a left or right delimiter (see section 4.6.3). The amsmath package defines the \lvert and \rvert commands for | as left and right delimiters, respectively, and \lVert and \rVert for \|, respectively.

4.6.2 Delimiters of fixed size

In this section, we'll discuss delimiters that do not change size; they are listed in Tables 4.1 and 4.2.

LaTeX knows that the symbols in these tables are delimiters, and spaces them accordingly. Notice the difference between $\|a\|$ and $\|a\|$. The first, $\|a\|$, was typed incorrectly: $\| a \|$. As a result, the vertical bars are too far apart. The second was typed correctly: $\|a\|$.

Delimiters are normally used in pairs but they can also be used singly. For instance, $F(x) |^{b}_{a}$ prints as $F(x)|^b_a$.

LaTeX provides the \bigl, \Bigl, \biggl, \Biggl, \bigr, \Bigr, \biggr, and \Biggr commands to produce big fixed-size delimiters. For example,

$($ $\bigl($ $\Bigl($ $\biggl($ $\Biggl($

prints:

(((((

For integral evaluation, you can choose one of the following:

$$F(x)|^b_a \quad F(x)\Big|^b_a \quad F(x)\bigg|^b_a$$

typed as

```
\[
   F(x) |^{b}_{a}          \quad
   F(x) \bigr|^{b}_{a}  \quad
   F(x) \Bigr|^{b}_{a}
\]
```

or \biggr, \Biggr for even larger symbols.

For fine-tuning, there is also \Big and \Bigg. For example,

$a\big|$ $a\Big|$ $a\Bigg|$

prints

$a\big| a\Big| a\Bigg|$

4.6.3 *Delimiters of variable size*

To enclose the formula

$$\frac{x^2 + 1}{a_2 b_2 \times c_2 d_2}$$

with parentheses of the appropriate size, for instance, in order to print

$$\left(\frac{x^2 + 1}{a_2 b_2 \times c_2 d_2} \right)^2$$

type

```
\[
   \left( \frac{x^2 + 1}{a_2 b_2 \times c_2 d_2} \right)^2
\]
```

The general construction is \left *delim1* and \right *delim2*, where *delim1* and *delim2* are chosen from Tables 4.1 and 4.2. LaTeX will inspect the formula and decide what size delimiters to use. Such delimiters *must be paired* in order for LaTeX to know the extent of the material to be vertically measured; however, the matching delimiters need not be the same. You sometimes need a *blank delimiter* to pair with some other delimiter. The left and right blank delimiters are designated by \left. and \right. respectively.

Here are some examples of delimiters of variable size:

$$\left|\frac{a+b}{2}\right|, \quad \|A^2\|, \quad \left(\frac{a}{2},b\right], \quad F(x)\big|_a^b$$

typed as:

```
\[
    \left| \frac{a + b}{2} \right|, \quad
    \left\| A^{2} \right\|,          \quad
    \left( \frac{a}{2}, b \right],   \quad
    \left. F(x) \right|_{a}^{b}
\]
```

The \left\langle command can be abbreviated as \left<.
The \right\rangle command can be abbreviated as \right>.

4.6.4   *Delimiters as binary relations*

The symbol | is used as a delimiter as well as a binary relation. As a binary relation it's written as \mid. For instance, $\{ x \mid x^2 \leq 2 \}$ is typed as

```
$\{\, x \mid x^{2} \leq 2 \,\}$
```

The \big and \bigg commands produce larger delimiters, spaced as symbols; \bigm and \biggm produce larger delimiters, spaced as binary relations. Example:

$$\left\{ x \,\middle|\, \int_0^x t^2\,dt \leq 5 \right\}$$

typed as

```
\[
    \left\{\, x \biggm| \int_{0}^x t^{2} \, dt \leq 5 \,\right\}
\]
```

4.7   *Operators*

You can't just type "sin x" for the sine function in math mode; for example,

```
$sin x$
```

prints $sin x$ instead of $\sin x$ as it should. The correct way to type this is

```
$\sin x$
```

The \sin command prints "sin" in the proper style and spacing. LaTeX calls \sin an *operator* (or "log-like function").

4.7.1 *Operator tables*

There are a number of operators such as \sin. They come in two varieties: simple ones, *operators without limits* (like \sin) and *operators with limits* (like \lim) that take a subscript in inline mode and a "limit" in displayed math mode.

Operators with limits inline example: $\lim_{x \to 0} f(x) = 1$, typed as

```
$\lim_{x \to 0}   f(x) = 1$
```

Operators with limits displayed example:

$$\lim_{x \to 0} f(x) = 1$$

typed as

```
\[
   \lim_{x \to 0} f(x) = 1
\]
```

The operators are listed in the Tables 4.3 and 4.4 (see also Appendix A). The \var commands in Table 4.3 and the \injlim and \projlim commands in Table 4.4 require the amsmath package.

The last four entries in Table 4.3 are illustrated by:

$$\varliminf_{x \to 0} \quad \varlimsup_{x \to 0} \quad \varinjlim_{x \to 0} \quad \varprojlim_{x \to 0}$$

which are typed as

```
\[
   \varliminf_{x \to 0} \quad  \varlimsup_{x \to 0}   \quad
   \varinjlim_{x \to 0} \quad  \varprojlim_{x \to 0}
\]
```

The following examples illustrate some entries from Table 4.4:

$$\injlim_{x \to 0} \quad \liminf_{x \to 0} \quad \limsup_{x \to 0} \quad \projlim_{x \to 0}$$

typed as:

\arccos	\arcsin	\arctan	\arg
\cos	\cosh	\cot	\coth
\csc	\dim	\exp	\hom
\ker	\lg	\ln	\log
\sec	\sin	\sinh	\tan
\tanh			
\varliminf	\varlimsup	\varinjlim	\varprojlim

Table 4.3: Operators without limits

\det	\gcd	\inf	\injlim
\lim	\liminf	\limsup	\max
\min	\projlim	\Pr	\sup

Table 4.4: Operators with limits

```
\[
    \injlim_{x \to 0} \quad \liminf_{x \to 0} \quad
    \limsup_{x \to 0} \quad \projlim_{x \to 0}
\]
```

You can force the limits in a displayed formula into the subscript position with the \nolimits command. Example:

$$\operatorname{inj\,lim}_{x \to 0} \quad \operatorname{lim\,inf}_{x \to 0} \quad \operatorname{lim\,sup}_{x \to 0} \quad \operatorname{proj\,lim}_{x \to 0}$$

typed as:

```
\[
    \injlim\nolimits_{x \to 0} \quad
    \liminf\nolimits_{x \to 0} \quad
    \limsup\nolimits_{x \to 0} \quad
    \projlim\nolimits_{x \to 0}
\]
```

4.7.2 *Declaring operators*

The amsopn package (automatically loaded by the amsmath package and, therefore, by all the $\mathcal{A}\mathcal{M}\mathcal{S}$-LATEX document classes) provides a command to define a new operator:

\DeclareMathOperator{*opCommand*}{*opName*}

You invoke the new operator with *opCommand* and the operator name to be typeset is *opName*. This command must be placed in the preamble. To define the operator Trunc, for example, issue the command

Type:	Print:
`$a \equiv v \mod{\theta}$`	$a \equiv v \mod \theta$
`$a \equiv v \bmod{\theta}$`	$a \equiv v \bmod \theta$
`$a \equiv v \pmod{\theta}$`	$a \equiv v \pmod \theta$
`$a \equiv v \pod{\theta}$`	$a \equiv v \pod \theta$

Table 4.5: Congruences

```
\DeclareMathOperator{\Trunc}{Trunc}
```

An operator is printed in math roman with a little space after it, so `$\Trunc A$` prints Trunc A.

The second argument is printed in math mode but – and * print as text. Here are some more examples; define

```
\DeclareMathOperator{\Trone}{Trunc_{1}}
\DeclareMathOperator{\Ststar}{Star-one*}
```

and then `$\Trone A$` prints $\mathrm{Trunc}_1 A$ and `$\Ststar A$` prints Star-one* A.

To define an operator with limits, use the *-ed form:

```
\DeclareMathOperator*{\Star}{Star}
```

and then

```
\[
    \Star_{x \to 0} \int_{0}^{x} f(x) \, dx = \pi
\]
```

will print

$$\Star_{x \to 0} \int_0^x f(x)\,dx = \pi$$

See also section 4.12.2.

4.7.3   Congruences

The "mod" function is a special operator used for congruences. Congruences are typeset with the \bmod and \pmod commands; if you use the amsopn package (automatically loaded by the amsmath package and by all the $\mathcal{AMS}$ document classes), then you get two additional variants, \mod and \pod, as shown in Table 4.5. The amsopn package slightly changes the LaTeX \pmod command: in inline math mode the spacing is reduced. See sections 9.1.2 and 9.4 for related user-defined commands.

Type:	Inline	Displayed	Type:	Inline	Displayed
`\prod_{i=1}^{n}`	$\prod_{i=1}^{n}$	$\prod\limits_{i=1}^{n}$	`\coprod_{i=1}^{n}`	$\coprod_{i=1}^{n}$	$\coprod\limits_{i=1}^{n}$
`\bigcap_{i=1}^{n}`	$\bigcap_{i=1}^{n}$	$\bigcap\limits_{i=1}^{n}$	`\bigcup_{i=1}^{n}`	$\bigcup_{i=1}^{n}$	$\bigcup\limits_{i=1}^{n}$
`\bigwedge_{i=1}^{n}`	$\bigwedge_{i=1}^{n}$	$\bigwedge\limits_{i=1}^{n}$	`\bigvee_{i=1}^{n}`	$\bigvee_{i=1}^{n}$	$\bigvee\limits_{i=1}^{n}$
`\bigsqcup_{i=1}^{n}`	$\bigsqcup_{i=1}^{n}$	$\bigsqcup\limits_{i=1}^{n}$	`\biguplus_{i=1}^{n}`	$\biguplus_{i=1}^{n}$	$\biguplus\limits_{i=1}^{n}$
`\bigotimes_{i=1}^{n}`	$\bigotimes_{i=1}^{n}$	$\bigotimes\limits_{i=1}^{n}$	`\bigoplus_{i=1}^{n}`	$\bigoplus_{i=1}^{n}$	$\bigoplus\limits_{i=1}^{n}$
`\bigodot_{i=1}^{n}`	$\bigodot_{i=1}^{n}$	$\bigodot\limits_{i=1}^{n}$	`\sum_{i=1}^{n}`	$\sum_{i=1}^{n}$	$\sum\limits_{i=1}^{n}$

Table 4.6: Large operators

4.8   Sums and products

Sums and products are special cases of a general construct called "large operators".

4.8.1   Large operators

Here is a sum in inline form $\sum_{i=1}^{n} x_i^2$ and also in displayed form

$$\sum_{i=1}^{n} x_i^2$$

In the latter form, the sum symbol is larger, and the subscript and superscript become *limits*.

Constructs that behave this way are called *large operators*. Table 4.6 gives a complete list of large operators.

Use the `\nolimits` command if you wish to show the limits of large operators as subscripts and superscripts in a displayed environment. For example,

$$\bigsqcup\nolimits_{\mathfrak{m}} X = a$$

is typed as

```
\[
    \bigsqcup\nolimits_{ \mathfrak{m} } X = a
\]
```

Sums and products are very important; so here are two more examples:

$$\frac{z^d - z_0^d}{z - z_0} = \sum_{k=1}^{d} z_0^{k-1} z^{d-k} \qquad (T^n)'(x_0) = \prod_{k=0}^{n-1} T'(x_k)$$

typed as

```
\[
    \frac{z^{d} - z_{0}^{d}}
        {z - z_{0}} =
    \sum_{k = 1}^{d} z_{0}^{k - 1} z^{d - k} \qquad
    (T^{n})'(x_{0}) = \prod_{k=0}^{n - 1} T'(x_{k})
\]
```

4.8.2   Multiline subscripts and superscripts

Large operators sometimes need multiline subscripts and superscripts. For this purpose, the amsmath package provides the \substack command.

For instance,

$$\sum_{\substack{i < n \\ i \text{ even}}} x_i^2$$

is typed as

```
\[
    \sum_{ \substack{ i < n\\
                      i \text{ even} } }
        x_{i}^{2}
\]
```

There is only one rule to remember: use the line separator \\. You can use the \substack command wherever subscripts or superscripts are used.

The lines are centered by substack; if you want them flush left, as in

$$\sum_{\substack{i < n \\ i \text{ even}}} x_i^2$$

then use the subarray environment with the argument l:

```
\[
    \sum_{ \begin{subarray}{l}
                i < n\\
                i \text{ even}
            \end{subarray} }
        x_{i}^{2}
\]
```

\hat{a}	$\hat{a}$	\Hat{a}	$\hat{a}$	\widehat{a}	$\widehat{a}$	a\sphat	$a^\wedge$
\tilde{a}	$\tilde{a}$	\Tilde{a}	$\tilde{a}$	\widetilde{a}	$\widetilde{a}$	a\sptilde	$a^\wedge$
\acute{a}	$\acute{a}$	\Acute{a}	$\acute{a}$				
\bar{a}	$\bar{a}$	\Bar{a}	$\bar{a}$				
\breve{a}	$\breve{a}$	\Breve{a}	$\breve{a}$			a\spbreve	$a^\smile$
\check{a}	$\check{a}$	\Check{a}	$\check{a}$			a\spcheck	$a^\vee$
\dot{a}	$\dot{a}$	\Dot{a}	$\dot{a}$			a\spdot	$a^\cdot$
\ddot{a}	$\ddot{a}$	\Ddot{a}	$\ddot{a}$			a\spddot	$a^{\cdot\cdot}$
\dddot{a}	$\dddot{a}$					a\spdddot	$a^{\cdot\cdot\cdot}$
\ddddot{a}	$\ddddot{a}$						
\grave{a}	$\grave{a}$	\Grave{a}	$\grave{a}$				
\vec{a}	$\vec{a}$	\Vec{a}	$\vec{a}$				

Table 4.7: Math accents

4.9 Math accents

The accents used in text (see section 2.4.6) can't be used in math; a separate set must be used. All math accents are shown in Table 4.7 (see also Appendix A).

The \dddot and \ddddot commands and all the capitalized commands in Table 4.7 require the amsmath package. The "sp" variety require the amsxtra package. Use the capitalized commands for double accents:

```
\[
    \Hat{\Hat{A}}\quad \hat{\hat{A}}
\]
```

prints

$$\hat{\hat{A}} \quad \hat{\hat{A}}$$

As you can see, \hat{\hat{A}} prints the double hat incorrectly.

The two "wide" varieties, \widehat and \widetilde, expand to fit their argument: $\widehat{A}$, $\widehat{ab}$, $\widehat{iii}$, $\widehat{aiai}$, $\widehat{iiiii}$, and $\widetilde{A}$, $\widetilde{ab}$, $\widetilde{iii}$, $\widetilde{aiai}$, $\widetilde{iiiii}$ (the latter is typed as $\widetilde{iiiii}$). If the "base" is too wide, the accent is centered:

$$\widehat{ABCDE}$$

The "sp" varieties, offered by the amsxtra package, are used for superscripts, as illustrated in Table 4.7. If you use a lot of accented characters, you'll appreciate user-defined commands (see section 9.1.1).

Notice the difference between $\bar{a}$ and $\overline{a}$, typed as

```
$\bar{a} \overline{a}$
```

For other examples of the \overline command, see section 4.10.2.

If your document uses a lot of double accented symbols, you can speed up the typesetting with the \accentedsymbol command provided by the amsxtra package. For instance, introduce

```
\accentedsymbol{\AHH}{\Hat{\Hat{A}}}
```

and invoke \Hat{\Hat{A}} with \AHH in math (or even better, define this command with \ensuremath so that you can also use it in text; see section 9.1.1). Note that the only difference between the above command and

```
\newcommand{\AHH}{\Hat{\Hat{A}}}
```

is the speed of execution. The price you pay for this increase in speed is that a symbol created with the \accentedsymbol command does not change size in subscripts and superscripts.

4.10   Horizontal lines that stretch

LaTeX and the amsmath package provide three types of "horizontal lines" above or below a formula that stretch: braces, bars, and arrows. There are also stretchable arrow math symbols.

4.10.1   Horizontal braces

The \overbrace command places a brace of variable size over its argument, as in

$$\overbrace{a + b + \cdots + z}$$

which is typed as

```
\[
    \overbrace{a + b + \cdots + z}
\]
```

A superscript adds a label to the brace, as in

$$\overbrace{a + a + \cdots + a}^{n}$$

This is typed as

```
\[
    \overbrace{a + a + \cdots + a}^{n}
\]
```

The \underbrace command works similarly, by placing a brace under the argument. A subscript adds a label to the brace, as in

$$\underbrace{a + a + \cdots + a}_{n}$$

which is typed as

```
\[
    \underbrace{a + a + \cdots + a}_{n}
\]
```

The following example combines these two commands:

$$\underbrace{\overbrace{a + \cdots + a}^{(m-n)/2} + \underbrace{b + \cdots + b}_{n} + \overbrace{a + \cdots + a}^{(m-n)/2}}_{m}$$

This is typed as

```
\[
    \underbrace{
        \overbrace{a + \cdots + a}^{(m - n)/2}
        + \underbrace{b + \cdots + b}_{n}
        + \overbrace{a + \cdots + a}^{(m - n)/2}
    }_{m}
\]
```

4.10.2   Over and underlines

You can overline and underline a formula with the \overline and \underline commands. Example:

$$\overline{X \cup \overline{\overline{X}}} = \overline{\overline{X}}$$

which is typed as

```
\[
    \overline{ \overline{X} \cup \overline{ \overline{X} } } =
    \overline{ \overline{X} }
\]
```

Similarly, you can place arrows over and under an expression:

$$\overleftarrow{a} \quad \overrightarrow{aa} \quad \overleftrightarrow{aaa} \quad \underleftarrow{aaaa} \quad \underrightarrow{aaaaa} \quad \underleftrightarrow{aaaaaa}$$

which is typed as

```
\[
    \overleftarrow{a}              \quad \overrightarrow{aa}         \quad
    \overleftrightarrow{aaa}  \quad \underleftarrow{aaaa}     \quad
    \underrightarrow{aaaaa}   \quad \underleftrightarrow{aaaaaa}
\]
```

The first two commands are standard LaTeX. The other four are provided by the amsmath package.

4.10.3   Stretchable arrow math symbols

The amsmath package provides a stretchable arrow math symbol that extends to accommodate a formula above the arrow with the \xleftarrow command, or below the arrow with the \xrightarrow command. The formula on top is given as the argument (possibly empty) and the formula below is an optional argument. Examples:

$$A \xrightarrow{1\cdot1} B \xleftarrow[\alpha \to \beta]{\text{onto}} C \xleftarrow[\gamma]{} D \leftarrow E$$

typed as

```
\[
    A \xrightarrow{\text{1-1}} B \xleftarrow[\alpha \to \beta]
    {\text{onto}} C \xleftarrow[\gamma]{} D \xleftarrow{} E
\]
```

4.11   The spacing of symbols

LaTeX provides a large variety of math symbols: Greek characters (α), binary operations ($\circ$), binary relations ($\leq$), negated binary relations ($\nleq$), arrows ($\nearrow$), delimiters ($\{$), and so on. All the math symbols are listed in the tables of Appendix A.

Consider the formula:

$$A = \{\, x \in X \mid x\beta \geq xy > (x+1)^2 - \alpha \,\}$$

typed as

```
\[
    A = \{\, x \in X \mid x \beta \geq x y
    > (x + 1)^{2} - \alpha \,\}
\]
```

The spacing of the symbols in the formula varies. In $x\beta$, the two symbols are very close. In $x \in X$, there is some space around $\in$, but in $x+1$, there is somewhat less

Short form:	Full form:	Size:	Short form:	Full form:
\,	\thinspace	␣	\!	\negthinspace
\:	\medspace	␣		\negmedspace
\;	\thickspace	␣		\negthickspace
	\quad	␣␣		
	\qquad	␣␣␣		

Table 4.8: Spacing commands

space around $+$. There is a little space after { and before } (not quite enough for this formula, which is why \, was added).

LaTeX spaces these symbols by classifying them into several categories. In the above formula, A, x, X, β, and so on, are *math symbols*; $=$, $\in$, $|$, $\geq$, and $>$ are *binary relations*; $+$ and $-$ are *binary operations*; and {, }, (, and) are *delimiters*.

As a rule, you do not have to be concerned with whether or not, say, $+$ is a binary operation in a formula. LaTeX knows it, and will typeset the formula correctly. However, in some situations, LaTeX does not know how to typeset a formula properly, and you'll have to give it a helping hand by adding spacing commands. Luckily, LaTeX and amsmath provide a variety of spacing commands, listed in Table 4.8. Six of these

\, \: \! \; \quad \qquad

are provided by LaTeX (however, the names \medspace and \thickspace are provided by the amsmath package), the other two by the amsmath package. The \neg commands remove space ("reverses" the print head).

The \quad and \qquad commands are normally used to adjust aligned formulas (see Chapter 5) or to add space before text in a math formula. The size of \quad and \qquad is dependent on the current font.

The \, and \! commands are most useful for fine-tuning math formulas. The amsmath package provides even finer control with the \mspace command and the math unit mu; 18 mu = 1 em (defined in section 2.8.3). Example: \mspace{3mu}.

The opening formula of this section is an important example of fine-tuning: in set notation, when using \mid for "such that", thin spaces are inserted just inside the braces. Here are some more examples of fine-tuning. (See also the formula gallery in section 1.3; there is one more example in section 4.12.1.)

Example 1   In section 1.2.3, you typed the formula $\int_0^\pi \sin x \, dx = 2$ as

$\int_{0}^{\pi} \sin x \, dx = 2$

You recall the spacing command "\," between \sin x and dx. Indeed, without it, LaTeX would have crowded $\sin x$ and dx: $\int_0^\pi \sin x dx = 2$

Example 2 $|-f(x)|$ (typed as $|-f(x)|$) is spaced incorrectly; LaTeX assumes that $|$ and f are regular math symbols, and $-$ is a binary operation. To get the correct spacing, type `$\left|-f(x)\right|$` which prints $|-f(x)|$; this form tells LaTeX that the first | is a left delimiter (see section 4.6), and therefore $-$ is the unary minus sign, not the binary subtraction operation. (See also section 4.6.1.)

Example 3   In $\sqrt{5}$side, typed as

`$\sqrt{5} \mbox{side}$`

$\sqrt{5}$ is too close to "side"; so type it as

`$\sqrt{5} \,\mbox{side}$`

which prints $\sqrt{5}$ side

Example 4   In $\sin x/\log n$, the division symbol / is too far from $\log n$, so type

`$\sin x / \! \log n$`

which prints $\sin x/\log n$

Example 5   In $f(1/\sqrt{n})$, typed as

`$f(1 / \sqrt{n})$`

the square root almost touches the closing parenthesis. To correct: $f(1/\sqrt{n}\,)$, type

`$f(1 / \sqrt{n}\,)$`

The symbol | in a math formula could be

- an ordinary math symbol, typed as |
- a binary relation, typed as `\mid`
- a left delimiter, typed as `\left|`
- a right delimiter, typed as `\right|`

Observe the spacing in $a|b$ and $a \mid b$ typed as `$a | b$` and `$a \mid b$`, respectively.

There is one more symbol with special spacing: the `\colon` command, used for formulas such as $f\colon A \to B$, typed as

`$f \colon A \to B$`

The colon (:) is a binary relation in math; `$f: A \to B$` prints $f : A \to B$.

See section 4.12.2 on how to declare the type of a symbol.

4.12   *Building new symbols*

LaTeX together with the AMSfonts font set provide a large variety of math symbols. But no matter how many symbols there are, users seem to want more. LaTeX and the amsmath package give you excellent tools to build new symbols from existing ones.

4.12.1   *Stacking symbols*

The LaTeX \stackrel{*top*}{*bottom*} command creates a new symbol by placing *top* (in the size of a superscript) above *bottom*. For instance, $x \overset{?}{=} y$ is typed as \stackrel{?}{=}. Note that a symbol created with \stackrel is always spaced as a binary relation, regardless of the original spacing rules for *bottom*.

The amsmath package lets you place any symbol *above*, or *below*, another. The \overset command has two arguments; the first argument is set in a smaller size *over* the second argument; the spacing rules of the second argument remain valid. The \underset command does the same but the first argument is set *under* the second argument. Examples:

$$\overset{\alpha}{a} \qquad \underset{\cdot}{X} \qquad \overset{\alpha}{a_i} \qquad \overset{\alpha}{a}_i$$

typed as

```
\[
    \overset{\alpha}{a}                    \qquad
    \underset{\boldsymbol{\cdot}}{X}       \qquad
    \overset{\alpha}{ a_{i} }              \qquad
    \overset{\alpha}{a}_{i}
\]
```

(For more information on the \boldsymbol command, see section 4.14.3.) Note that in the third example, $\overset{\alpha}{a_i}$, the α seems to be sitting too far to the right; the fourth example corrects that.

You can use these commands for binary relations, as in

$$f(x) \overset{\text{def}}{=} x^2 - 1$$

typed as

```
\[
    f(x) \overset{ \text{def} }{=} x^{2} - 1
\]
```

Note that $\overset{\text{def}}{=}$ remains a binary relation, as witnessed by the spacing on either side. Here is another example:

$$\frac{a}{b} \overset{u}{+} \frac{c}{d} \overset{l}{+} \frac{e}{f}$$

typed as

```
\[
    \frac{a}{b} \overset{u}{+} \frac{c}{d}
    \overset{l}{+} \frac{e}{f}
\]
```

Note that $\overset{u}{+}$ and $\overset{l}{+}$ are properly spaced as binary operations.

You can *negate* a symbol with the \not command; for instance, $a \notin b$ and $a \neq b$ are typed as $a \not\in b$ and $a \not= b$, respectively. It is preferable, however, to use the negated symbols, $\notin$ (an $\mathcal{AMS}$ symbol), typed as $\notin$, and $\neq$, typed as $\ne$ (see the $\mathcal{AMS}$ negated binary relations table in Appendix A). For instance, "*a* does not divide *b*", $a \nmid b$, should be typed as $a \nmid b$, not as $a \not\mid b$, which prints $a \not\mid b$. (In section 4.12.2, you'll learn how to do $a \not\mid b$ a bit better, namely, $a \not\mid b$, typed as $a \mathrel{\not|} b$. However, $a \nmid b$ is still best.)

Finally, there is the \sideset command with three arguments and the strange form

\sideset{ _{ll}^{ul} }{ _{lr}^{ur} }{large_op}

where ll stands for the symbol to be placed at the lower left, ul for upper left, lr for lower right, and ur for upper right; $large\_op$ is a large operator. Examples:

$$\prod_{*}^{*} \text{ and } {}^{*}\prod$$

typed as

\[
    \sideset{}{_{*}^{*}}{\prod}\text{ and }\sideset{^{*}}{}{\prod}
\]

Here is a more meaningful example:

\[
    \sideset{}{'}{\sum}_{\substack{ i < 10\\ j < 10 } } x_{i} z_{j}
\]

which prints

$$\sideset{}{'}{\sum}_{\substack{i<10\\j<10}} x_i z_j$$

In this example note that ' is an accented symbol, so you do not have to type ^' in the second argument. Typing \sum' would not work, since it would place the prime on top of the sum symbol.

4.12.2   *Declaring the type*

You have seen in section 4.11 that some symbols are binary relations and some are binary operations. In fact, you can declare any symbol to be either. The \mathbin command declares its argument to be a binary operation, for instance,

```
\mathbin{\alpha}
```

makes this instance of \alpha behave like a binary operation, as in $a \alpha b$, typed as

```
$a \mathbin{\alpha} b$
```

Use the \mathrel command to declare a binary relation, as in

```
\mathrel{ \text{fine} }
```

Then in the formula a fine b, typed as

```
$a \mathrel{ \text{fine} } b$
```

where the string "fine" is handled as a binary relation.

If you use the amsopn package (or amsmath, which loads amsopn), then anything can be declared an operator or an operator with limits. For instance, to use Trunc as an operator, type

```
$\operatorname{Trunc} f(x)$
```

which will print: Trunc $f(x)$.

To use Trunc as an operator with limits, type

```
\[
    \operatorname*{Trunc}_{x \in X} A_{x}
\]
```

which prints:

$$\operatorname*{Trunc}_{x \in X} A_x$$

Typesetting the same formula inline gives $\operatorname{Trunc}_{x \in X} A_x$.

In section 4.7.2, we discussed the alternative approach, provided by the amsopn package, namely, the \DeclareMathOperator and \DeclareMathOperator* commands.

4.13 *Vertical spacing*

As a rule, all horizontal and vertical spacing in a math formula is done by LaTeX. Nevertheless, you have to adjust horizontal spacing quite often (see section 4.11). In contrast, there is seldom a need to adjust the vertical spacing, but there are a few exceptions.

The formula $\sqrt{a} + \sqrt{b}$ does not look right, since the square roots are not uniform in size. You can correct this with \mathstrut, which inserts an invisible vertical space:

```
$\sqrt{\mathstrut a} + \sqrt{\mathstrut b}$
```

which prints: $\sqrt{a} + \sqrt{b}$. See also section 2.9.4 for more information on struts.

Another way to handle this situation is with the vertical phantom command, \vphantom, which measures the height of its argument and places a mathstrut of this height into the formula. So

$\sqrt{\vphantom{b} a} + \sqrt{b}$

also prints uniform square roots, $\sqrt{a} + \sqrt{b}$, but is more versatile than the previous solution.

The \smash command directs LaTeX to pretend that its argument does not reach above or below the line in which it is typeset.

For instance, the two lines of the admonition:

It is **very important** that you memorize the integral $\frac{1}{\int f(x)\,dx} = 2g(x) + C$, which will appear on the next test.

are too far apart; LaTeX had to make room for the fraction. However, in this instance, it's not necessary because the second line is very short. So redo the two lines:

```
It is \textbf{very important} that you memorize the integral
$\smash{\frac{1}{\int f(x) \, dx}} = 2 g(x) + C$, which will
appear on the next test.
```

which prints:

It is **very important** that you memorize the integral $\frac{1}{\int f(x)\,dx} = 2g(x) + C$, which will appear on the next test.

With the amsmath package, there is a \smash[t] command to smash the top, and a \smash[b] command to smash the bottom.

4.14   *Math alphabets and symbols*

The classification of math symbols in the context of spacing was discussed in section 4.11. The symbols in a formula can also be classified as *characters from math alphabets* and *math symbols*. In the formula

$$A = \{x \in X \mid x\beta \geq xy > (x+1)^2 - \alpha\}$$

the following characters come from a math alphabet:

$$A \quad x \quad X \quad y \quad 1 \quad 2$$

while these are regarded as math symbols:

$$= \quad \{ \quad \in \quad | \quad \beta \quad \geq \quad > \quad (\quad + \quad) \quad - \quad \alpha \quad \}$$

4.14.1 *Math alphabets*

The letters and digits you type in a math formula come from a *math alphabet*. The *default math alphabet*—the one you get if you do not ask for something else—is Computer Modern math italic for *letters*. In the formula $x^2 \vee y_3 = \alpha$, x and y come from this math alphabet. The *default math alphabet for digits* is Computer Modern roman; 2 and 3 in this formula are typeset in Computer Modern roman.

LaTeX has a number of commands to switch to various math alphabets; two of them, which select text fonts as math alphabets, are the most important:

Command:	Math alphabet:	Produces:
\mathbf{a}	math bold	a
\mathit{a}	math italic	a

As a rule, a command that switches to a math alphabet should be used with *a single-letter argument*. Some users make an exception, recommending that \mathit be used with word identifiers, as in the example below:

$$credit + debit = 0$$

typed as

```
\[
    \mathit{credit} + \mathit{debit} = 0
\]
```

Beware of the pitfalls, however; for instance, \mathit{left-side} will be typeset as $left - side$, printing a minus for the hyphen.

There are four more commands that switch to math alphabets:

Command:	Math Alphabet:	Produces:
\mathsf{a}	math sans serif	a
\mathrm{a}	math roman	a
\mathtt{a}	math typewriter	a
\mathnormal{a}	math italic	a

Roman is used in formulas either for operator names (such as "sin" in $\sin x$) or for text. For operator names, you should use the command \operatorname or \DeclareMathOperator (or the *-ed version), which sets the name of the operator in roman, and also provides the proper spacing (see section 4.12.2). For text, you should use the \mbox command (or the \text command if you use the amsmath package—see section 4.5).

The \mathnormal command switches to the default math alphabet, but this command is seldom used in practice.

The Computer Modern fonts provide one more useful math alphabet, but LaTeX and the amsmath package do not define a command to access it. It is the math bold italic alphabet, as in *a **A** b **B***. In section 9.5 (see page 296), you'll find a user-defined command to utilize it.

4.14.2   *Math alphabets of symbols*

At the beginning of this section, you may have noticed that α was not classified as belonging to an alphabet. Indeed, α is treated by LaTeX as a math symbol rather than a member of a math alphabet. You can't italicize it or slant it, nor is there a sans serif version. There is a bold version (the AMSFonts even has it in smaller sizes) but you must use the \boldsymbol command in the amsbsy package (automatically loaded by all the $\mathcal{AMS}$-LaTeX document classes) to produce it.

Two "alphabets of symbols" are built into LaTeX: the Greek alphabet (see section A-1 for the symbol table) and calligraphic, an uppercase-only alphabet, invoked by the \mathcal command; for example, $\mathcal{A, C, E}$ is typed as

```
$\mathcal{A}$, $\mathcal{C}$, $\mathcal{E}$
```

The eucal package (see section 8.5) provides the Euler Script alphabet as a substitute for the calligraphic alphabet. Type

```
\usepackage{eucal}
```

in the preamble to *redefine* the \mathcal command. With this package,

```
$\mathcal{A}$, $\mathcal{C}$, $\mathcal{E}$
```

now print as $\mathcal{A}, \mathcal{C}, \mathcal{E}$. However, if you invoke the eucal package with the option:

```
\usepackage[mathscr]{eucal}
```

then the \mathcal command will continue to provide the calligraphic alphabet and the \mathscr command will produce the Euler Script alphabet:

$\mathcal{A, C, E}$
$\mathscr{A, C, E}$

is typed as

```
$\mathcal{A}$, $\mathcal{C}$, $\mathcal{E}$
```

```
$\mathscr{A}$, $\mathscr{C}$, $\mathscr{E}$
```

In the preamble, the

```
\usepackage{eufrak}
```

command invokes the eufrak package (see section 8.5), which gives the \mathfrak command for Euler Fraktur, for both uppercase and lowercase characters. The sample characters n, p, 𝔑, 𝔓 are typed as

```
$\mathfrak{n}$, $\mathfrak{p}$, $\mathfrak{N}$, $\mathfrak{P}$
```

If you use the amsfonts package, then you do not have to invoke the eufrak package to use Euler Fraktur.

Blackboard bold, as in 𝔸, 𝔹, ℂ, is provided in the amsfonts package with the \mathbb command. It is also an uppercase-only math alphabet. The above sample was typed as

```
$\mathbb{A}$, $\mathbb{B}$, $\mathbb{C}$
```

4.14.3 Bold math symbols

In math, most of the font characteristics are specified by LATEX. One exception is bold face. To make a *letter* (from a math alphabet) bold in a formula, use the \mathbf command. For instance, in

let the vector **v** be chosen ...

the bold "v" is produced by $\mathbf{v}$.

To obtain *bold math symbols*, use the \boldsymbol command provided by the amsbsy (𝒜𝑀𝒮 bold symbols) package, automatically loaded by all the 𝒜𝑀𝒮-LATEX document classes and by the amsmath package; for instance,

$$5 \quad \boldsymbol{\alpha} \quad \boldsymbol{\Lambda} \quad \boldsymbol{\mathcal{A}} \quad \rightarrow \quad \boldsymbol{A}$$

is typed as

```
\[
    \boldsymbol{5}          \quad \boldsymbol{\alpha}          \quad
    \boldsymbol{\Lambda} \quad \boldsymbol{ \mathcal{A} } \quad
    \boldsymbol{\to}       \quad \boldsymbol{A}
\]
```

Note that \boldsymbol{A} prints A, bold math italic A.

If you do not have AMSFonts installed, many bold symbols presented in this book will not be available.

To make an entire formula bold, use the \mathversion{bold} command as in

```
{\mathversion{bold} $a \equiv c \pmod{\theta}$}
```

which prints: $a \equiv c \pmod{\boldsymbol{\theta}}$. Note that the \mathversion{bold} command is given *before the formula*.

So to typeset $\boldsymbol{\mathcal{AMS}}$, type

```
$\boldsymbol{ \mathcal{A} } \boldsymbol{ \mathcal{M} }
   \boldsymbol{ \mathcal{S} }$
```

or

```
$\boldsymbol{ \mathcal{AMS} }
```

or

```
{\mathversion{bold} $\mathcal{AMS}$}
```

Within the scope of \mathversion{bold}, you can undo its effect with

```
\mathversion{normal}
```

Not all symbols have bold variants; if you type

```
$\sum \quad \boldsymbol{\sum}$
```

you get: $\sum \quad \sum$ (that is, the two are identical). To obtain a bold version, use the *poor man's bold* invoked with the \pmb command provided by the amsbsy package (automatically loaded by the amsmath package). This prints the symbol three times very close to each other. For some symbols the result is satisfactory. However, \pmb destroys the "type" of the symbol: \pmb{\sum} is no longer a large operator. To make it into a large operator, declare it:

```
\mathop{\pmb{\sum}}
```

Compare the following three variants of sum:

$$\sum_{i=1}^{n} i^2 \quad \sum_{i=1}^{n} i^2 \quad \sum_{i=1}^{n} i^2$$

The first sum is typed (in displayed math mode) as

```
\sum_{i = 1}^{n} i^{2}
```

The second uses poor man's bold, but does not declare the result a large operator:

```
\pmb{\sum}_{i = 1}^{n} i^{2}
```

The third uses poor man's bold and declares the result to be a large operator:

```
\mathop{\pmb{\sum}}_{i = 1}^{n} i^{2}
```

4.14.4 *Size changes*

There are four math sizes, invoked by the command declarations:

- \displaystyle, the normal size for displayed formulas
- \textstyle, the normal size for inline formulas
- \scriptstyle, the normal size for subscripted and superscripted symbols
- \scriptscriptstyle, the normal size for doubly subscripted and superscripted symbols

These commands control a number of style parameters in addition to the size. For example, to obtain the continued fraction

$$\cfrac{1}{2 + \cfrac{1}{3 + \cdots}}$$

type

```
\[
    \frac{1}{\displaystyle 2 + \frac{1}{\displaystyle 3 + \cdots}}
\]
```

4.14.5 *Continued fractions*

The amsmath package makes typesetting continued fractions, such as the example in section 4.14.4, easier with the command \cfrac:

```
\[
    \cfrac{1}{ 2 + \cfrac{1}{3 + \cdots} }
\]
```

For comparison, look at this formula typed with \frac:

$$\frac{1}{2 + \frac{1}{3 + \cdots}}$$

Use \cfrac[l] (resp., \cfrac[r]) to place the numerator on the left (resp., right). For instance,

$$\cfrac[l]{1}{2 + \cfrac[l]{1}{3 + \cdots}}$$

is typed as

```
\[
    \cfrac[l]{1}{ 2 + \cfrac[l]{1}{3 + \cdots} }
\]
```

4.15   Tagging and grouping

In addition to numbering, the amsmath package also allows the \tag-ing of equations. In the equation and equation* environments,

\tag{name}

attaches the "tag" name to the equation; the tag replaces the number.

Recall that the numbering of an equation is *relative*; that is, the number assigned is relative to the position of the equation with respect to other equations in the document. The tagging of an equation (with text), on the other hand, is *absolute*; the tag remains the same even after rearrangement. A numbered equation needs a \label{*name*}, so \ref{*name*} can reference the number generated by LaTeX. (An equation may contain both a tag and a label. The tag will print; the label can be used for page reference with the \pageref command—see section 6.4.2.)

Note that if there is a tag, the equation and the equation* environments are equivalent. Example:

(Int) $$\int_{-\infty}^{\infty} e^{-x^2}\, dx = \sqrt{\pi}$$

may also be typed as

\begin{equation*}
    \int_{-\infty}^{\infty} e^{-x^{2}} \, dx = \sqrt{\pi}\tag{Int}
\end{equation*}

The \tag* command (also in the amsmath package) is the same as \tag except that it does not automatically use parentheses. To get

A–B $$\int_{-\infty}^{\infty} e^{-x^2}\, dx = \sqrt{\pi}$$

type

\begin{equation}
    \int_{-\infty}^{\infty} e^{-x^{2}} \, dx = \sqrt{\pi}
        \tag*{A--B}
\end{equation}

Tagging allows numbered "variants" of equations. For instance, you may have the equation

(1) $$A^{[2]} \diamond B^{[2]} \cong (A \diamond B)^{[2]}$$

and introduce a variant (*anywhere* else in the document)

(1′) $$A^{\langle 2 \rangle} \diamond B^{\langle 2 \rangle} \cong (A \diamond B)^{\langle 2 \rangle}$$

which you want to number the same as the original but primed. If the label of the first equation is E:first, then the second equation is typed as follows:

```
\begin{equation} \tag{\ref{E:first}$'$}
   A^{\langle 2 \rangle} \diamond B^{\langle 2\rangle}
   \cong (A \diamond B)^{\langle 2 \rangle}
\end{equation}
```

Such a tag is not absolute. It changes with the referenced label; it does not change if the tagged equation is moved, however.

In contrast, *grouping* (also provided by the amsmath package) applies to a group of *adjacent* equations. Suppose the last equation was numbered (1) and you want the next group of equations referred to as (2), with individual equations numbered as (2a), (2b), and so on. To do this, bracket these equations in a subequations environment. For instance:

$$(2a) \qquad\qquad A^{[2]} \diamond B^{[2]} \cong (A \diamond B)^{[2]}$$

followed by the variant:

$$(2b) \qquad\qquad A^{\langle 2 \rangle} \diamond B^{\langle 2 \rangle} \cong (A \diamond B)^{\langle 2 \rangle}$$

is typed as

```
\begin{subequations} \label{E:joint}
   \begin{equation} \label{E:original}
      A^{[2]} \diamond B^{[2]} \cong (A \diamond B)^{[2]}
   \end{equation}
   followed by the variant:
   \begin{equation} \label{E:modified}
      A^{\langle 2 \rangle} \diamond B^{\langle 2 \rangle}
      \cong (A \diamond B)^{\langle 2 \rangle}
   \end{equation}
\end{subequations}
```

and then

\eqref{E:joint} prints (2)

\eqref{E:original} prints (2a)

\eqref{E:modified} prints (2b)

Note that in this example, references to the second and third labels produce numbers that also appear in the typeset version. The group label, \label{E:joint}, is there for reference to the entire group.

The subequations environments can contain the multiline math constructs of Chapter 5. For a simple example, see section 5.3.1.

4.16   Generalized fractions

The amsmath package provides a generalized fraction command:

\genfrac{*left-delim*}{*right-delim*}{*thickness*}{*mathstyle*}
        {*numerator*}{*denominator*}

where

- *left-delim* is the left delimiter for the formula; default: no left delimiter;
- *right-delim* is the right delimiter for the formula; default: no right delimiter;
- *thickness* is the thickness of the fraction line; default: the normal weight;
- *mathstyle* is

    0 for \displaystyle,

    1 for \textstyle,

    2 for \scriptstyle,

    3 for \scriptscriptstyle,

    default: depends on the context; if the formula is being set in display style, then the default is 0, and so on;

- *numerator* is the numerator;
- *denominator* is the denominator.

Examples: \frac{*numerator*}{*denominator*} is the same as

\genfrac{}{}{}{}{*numerator*}{*denominator*}

\dfrac{*numerator*}{*denominator*} is the same as

\genfrac{}{}{}{0}{*numerator*}{*denominator*}

\tfrac{*numerator*}{*denominator*} is the same as

\genfrac{}{}{}{1}{*numerator*}{*denominator*}

\binom{*numerator*}{*denominator*} is the same as

\genfrac{(}{)}{0pt}{}{*numerator*}{*denominator*}

Here are some more examples:

$$\frac{a+b}{c} \quad \frac{a+b}{c} \quad \frac{a+b}{c} \quad \frac{a+b}{c} \quad \left[\frac{a+b}{c}\right] \quad \left]\frac{a+b}{c}\right[$$

typed as

```
\[
    \frac{a + b}{c} \quad
    \genfrac{}{}{1pt}{}{a + b}{c}    \quad
    \genfrac{}{}{1.5pt}{}{a + b}{c}  \quad
    \genfrac{}{}{2pt}{}{a + b}{c}    \quad
    \genfrac{[}{]}{0pt}{}{a + b}{c}  \quad
    \genfrac{]}{[}{0pt}{}{a + b}{c}
\]
```

The delimiters must be chosen from Tables 4.1 and 4.2.

If you repeatedly use a \genfrac construct, name it; see section 9.1 for user-defined commands.

4.17 Boxed formulas

The \boxed command puts its argument in a box, as in

$$\boxed{\int_{-\infty}^{\infty} e^{-x^2}\,dx = \sqrt{\pi}}$$

typed as

```
\[
    \boxed{ \int_{-\infty}^{\infty} e^{-x^{2}} \, dx = \sqrt{\pi} }
\]
```

This command can also be used with a text argument.

Multiline math displays

This is, possibly, the most important chapter on $\mathcal{AMS}$-LaTeX: a discussion of how to display math formulas on more than one line using the constructs provided by the amsmath package. See the inside back cover of this book for thumbnail sketches of some of the basic constructs. For most mathematics articles, three constructs will suffice: *simple align*, *annotated align* (which are specials cases of align), and cases, introduced in section 1.4.2.

Multiline math formulas are displayed in *columns*, which are either *adjusted* (centered, or flush left or flush right) or *aligned* (an alignment point is designated for each column in each line).

You'll start with two adjusted, one-column math environments: gather and multline. In section 5.3, some general rules are stated that apply to all multiline math environments, and group numbering is discussed.

The align environment is one of the major contributions of the amsmath package; as you have seen in section 1.4.2, it's used to create aligned formulas. In section 5.4, you learn how to use it for more complicated aligned multicolumn constructs.

While math environments like gather, multline, and align produce complete displays, *subsidiary math environments* create a "large symbol" that may be used in any math formula. The *aligned subsidiary* math environments (`aligned` and `gathered`) are discussed in section 5.5, while the *adjusted subsidiary* math environments (`matrix` and `array`) are presented in section 5.6.

Section 5.7 discusses a subsidiary math environment that typesets simple commutative diagrams. Finally, section 5.8 describes how to allow pagebreaks in multiline math environments.

All the constructs of this chapter except the `array` and the `eqnarray` environments require the amsmath package; you'll not be reminded of this in the text. The document `multline.tpl` in the `ftp` directory (see page 4) contains all the multiline formulas of this chapter. See section 8.9.2 of *The LaTeX Companion* and [5] (and specifically, section 14.6 of `amsmath.dtx`) for the style parameters controlling these constructs.

5.1 *Gathering formulas*

Gathering a number of one-line formulas, each centered on a line, is provided by the gather environment; for instance:

$$(1) \qquad x_1 x_2 + x_1^2 x_2^2 + x_3$$
$$(2) \qquad x_1 x_3 + x_1^2 x_3^2 + x_2$$
$$(3) \qquad x_1 x_2 x_3$$

Formulas (1)–(3) are typed as follows

```
\begin{gather}
  x_{1} x_{2} + x_{1}^{2} x_{2}^{2} + x_{3} \label{E:mm1.1}\\
  x_{1} x_{3} + x_{1}^{2} x_{3}^{2} + x_{2} \label{E:mm1.2}\\
  x_{1} x_{2} x_{3} \label{E:mm1.3}
\end{gather}
```

Rule ■ gather environment

1. \\ separates the lines; there is no \\ at the end of the last line.
2. Each line is numbered except those that are \tag-ed and those whose numbering is prohibited by \notag.
3. No blank line is permitted in the environment.

The gather* environment is like gather, except that all lines are unnumbered (but can be \tag-ed).

5.2 *Splitting a long formula*

The multline environment is used to split one very long formula into several lines; the first line is flush left, the last is flush right, and the middle lines are centered:

$$
(4) \quad (x_1 x_2 x_3 x_4 x_5 x_6)^2
$$
$$
+ (x_1 x_2 x_3 x_4 x_5 + x_1 x_3 x_4 x_5 x_6 + x_1 x_2 x_4 x_5 x_6 + x_1 x_2 x_3 x_5 x_6)^2
$$
$$
+ (x_1 x_2 x_3 x_4 + x_1 x_2 x_3 x_5 + x_1 x_2 x_4 x_5 + x_1 x_3 x_4 x_5)^2
$$

This is typed as:

```
\begin{multline}\label{E:mm2}
   (x_{1} x_{2} x_{3} x_{4} x_{5} x_{6})^{2}\\
   + (x_{1} x_{2} x_{3} x_{4} x_{5}
    + x_{1} x_{3} x_{4} x_{5} x_{6}
    + x_{1} x_{2} x_{4} x_{5} x_{6}
    + x_{1} x_{2} x_{3} x_{5} x_{6})^{2}\\
   + (x_{1} x_{2} x_{3} x_{4} + x_{1} x_{2} x_{3} x_{5}
    + x_{1} x_{2} x_{4} x_{5} + x_{1} x_{3} x_{4} x_{5})^{2}
\end{multline}
```

Rule ■ multline environment

1. \\ separates the lines; there is no \\ at the end of the last line.
2. The whole formula is numbered unless it's \tag-ed or the numbering is suppressed with \notag.
3. No blank line is permitted in the environment.

There is one more rule that we discuss in section 5.3.1.

A typical error is to write "multiline" for "multline", giving the message:

```
! LaTeX Error: Environment multiline undefined.
```

```
l.5 \begin{multiline}
                    \label{E:mm2}
```

In the multline* environment, all lines are unnumbered (but can be jointly \tag-ed).

The indentation of the first and last lines from the margin is controlled by the \multlinegap length command (with a default of 10 pt) unless there is a tag on one of those lines. This can be reset, witness:

```
\begin{multline*}
  (x_{1} x_{2} x_{3} x_{4} x_{5} x_{6})^{2}\\
  + (x_{1} x_{2} x_{3} x_{4} x_{5}
  + x_{1} x_{3} x_{4} x_{5} x_{6}
  + x_{1} x_{2} x_{4} x_{5} x_{6}
  + x_{1} x_{2} x_{3} x_{5} x_{6})^{2}\\
  + (x_{1} x_{2} x_{3} x_{4} + x_{1} x_{2} x_{3} x_{5}
  + x_{1} x_{2} x_{4} x_{5} + x_{1} x_{3} x_{4} x_{5})^{2}
\end{multline*}
\begin{setlength}{\multlinegap}{0pt}
  \begin{multline*}
    (x_{1} x_{2} x_{3} x_{4} x_{5} x_{6})^{2}\\
    + (x_{1} x_{2} x_{3} x_{4} x_{5}
    + x_{1} x_{3} x_{4} x_{5} x_{6}
    + x_{1} x_{2} x_{4} x_{5} x_{6}
    + x_{1} x_{2} x_{3} x_{5} x_{6})^{2}\\
    + (x_{1} x_{2} x_{3} x_{4} + x_{1} x_{2} x_{3} x_{5}
    + x_{1} x_{2} x_{4} x_{5} + x_{1} x_{3} x_{4} x_{5})^{2}
  \end{multline*}
\end{setlength}
```

which prints

$$(x_1x_2x_3x_4x_5x_6)^2$$
$$+ (x_1x_2x_3x_4x_5 + x_1x_3x_4x_5x_6 + x_1x_2x_4x_5x_6 + x_1x_2x_3x_5x_6)^2$$
$$+ (x_1x_2x_3x_4 + x_1x_2x_3x_5 + x_1x_2x_4x_5 + x_1x_3x_4x_5)^2$$

$$(x_1x_2x_3x_4x_5x_6)^2$$
$$+ (x_1x_2x_3x_4x_5 + x_1x_3x_4x_5x_6 + x_1x_2x_4x_5x_6 + x_1x_2x_3x_5x_6)^2$$
$$+ (x_1x_2x_3x_4 + x_1x_2x_3x_5 + x_1x_2x_4x_5 + x_1x_3x_4x_5)^2$$

A single line of a `multline` (or `multline*`) environment can be typeset flush left or flush right with the `\shoveleft` and `\shoveright` commands, respectively; for instance, to typeset the second line of formula (4) flush left:

$$(x_1x_2x_3x_4x_5x_6)^2$$
$$+ (x_1x_2x_3x_4x_5 + x_1x_3x_4x_5x_6 + x_1x_2x_4x_5x_6 + x_1x_2x_3x_5x_6)^2$$
$$+ (x_1x_2x_3x_4 + x_1x_2x_3x_5 + x_1x_2x_4x_5 + x_1x_3x_4x_5)^2$$

type it as follows

```
\begin{multline*}
   (x_{1} x_{2} x_{3} x_{4} x_{5} x_{6})^{2}\\
   \shoveleft{+ (x_{1} x_{2} x_{3} x_{4} x_{5}
   + x_{1} x_{3} x_{4} x_{5} x_{6}
   + x_{1} x_{2} x_{4} x_{5} x_{6}
   + x_{1} x_{2} x_{3} x_{5} x_{6})^{2}}\\
 + (x_{1} x_{2} x_{3} x_{4} + x_{1} x_{2} x_{3} x_{5}
   + x_{1} x_{2} x_{4} x_{5} + x_{1} x_{3} x_{4} x_{5})^{2}
\end{multline*}
```

Observe that the entire line constitutes the argument of the \shoveleft command, which is followed by \\ (unless it's the last line).

5.3   Some general rules

Even though you have only seen a few examples of multiline math environments, I should point out that the multiline math environments and subsidiary environments share a number of rules. A few are listed below.

Rule ■ Multiline math environments

1. \\ separates the lines; there is no \\ at the end of the last line.
2. No blank line is permitted in the environment.
3. If the environment contains a number of formulas, then, as a rule, each formula is numbered. If you add a \label, then the equation number generated can be cross-referenced.
4. You can override the numbering with a \notag command.
5. You can also override the numbering with the \tag command, which works just as it does for equations (see section 4.15).
6. \tag and \label should precede the line separator \\except on the last line. There can be no \tag outside the environment.
7. For cross-referencing, use \label and \ref the same way as you do for an equation (see section 6.4.2).
8. Each multiline math environment has a *-ed form, where each formula becomes an *unnumbered equation*, that is, all lines are unnumbered (but can be \tag-ed).

A \notag placed after the environment is ignored, while a \tag gives the error message:

```
! Package amsmath Error: \tag not allowed here.
```

5.3.1   *The subformula rule*

The formula in the `multline` environment is split up into a number of parts by the \\ symbols; for instance, formula (4) is split into three parts:

```
(x_{1} x_{2} x_{3} x_{4} x_{5} x_{6})^{2}
```

and

```
  + (x_{1} x_{2} x_{3} x_{4} x_{5}
   + x_{1} x_{3} x_{4} x_{5} x_{6}
   + x_{1} x_{2} x_{4} x_{5} x_{6}
   + x_{1} x_{2} x_{3} x_{5} x_{6})^{2}
```

and

```
  + (x_{1} x_{2} x_{3} x_{4} + x_{1} x_{2} x_{3} x_{5} +
   x_{1} x_{2} x_{4} x_{5} + x_{1} x_{3} x_{4} x_{5})^{2}
```

Such parts of a formula are called *subformulas*.

The aligned formula $x = y+z$, (in the simple align example—see section 1.4.2) typed as

```
x &= y + z,
```

splits up into two parts:

```
x
```

and

```
= y + z,
```

In general, the first part is from the beginning of the formula to the first & symbol. There can be any number of parts delimited by two consecutive & symbols. Finally, the last part is from the last & symbol to the end of the formula. These are also called *subformulas*.

And here is the last of the general rules:

Rule ■ Subformulas

A subformula must be a formula that LaTeX can typeset.

Suppose that you want to split the formula

$$x_1 + y_1 + \left(\sum_{i<5} \binom{5}{i} + a^2 \right)^2$$

just before the binomial coefficient. Try

```
\begin{multline}
   x_{1} + y_{1} + \left( \sum_{i < 5}\\
   \binom{4}{i} + a^{2} \right)^{2}
\end{multline}
```

When typesetting this formula, you'll get the error message:

```
! Missing \right. inserted.
<inserted text>
                \right .
...
l.11 \end{multline}
```

Of course, the first subformula violated the subformula rule since

```
x_{1} + y_{1} + \left( \sum_{i < 5}
```

is a subformula that LaTeX can't typeset (the `\left(` command must be matched by the `\right` command and some delimiter).

Testing for the subformula rule is easy. Split up the formula into its subformulas, and try to typeset each subformula separately.

5.3.2   *Group numbering*

In most constructs in this chapter, you'll obtain a number of equations typeset together, arranged in some way (aligned or adjusted). Each equation is numbered separately, unless `\tag`-ed or `\notag`-ed. Often, you may want the equations to share a common number, but would want to reference each equation separately.

How do you change the numbering of the equations in formulas (1)–(3) (on page 181) to (1), (1a), and (1b)? This can be done as follows:

```
\begin{gather}
   x_{1} x_{2} + x_{1}^{2} x_{2}^{2} + x_{3} \label{E:mm1}        \\
   x_{1} x_{3} + x_{1}^{2} x_{3}^{2} + x_{2} \tag{\ref{E:mm1}a}\\
   x_{1} x_{2} x_{3} \tag{\ref{E:mm1}b}
\end{gather}
```

which prints

$$(1) \qquad\qquad x_1 x_2 + x_1^2 x_2^2 + x_3$$

$$(1a) \qquad\qquad x_1 x_3 + x_1^2 x_3^2 + x_2$$

$$(1b) \qquad\qquad x_1 x_2 x_3$$

To obtain $(1')$, type

```
\tag{\ref{E:mm1}$'$}
```

and for (1_a), type

`\tag{\ref{E:mm1}${}_{\text{a}}$}`

Alternatively, you may include the gather environment in a subequations environment (see section 4.15):

$$
\begin{align*}
\text{(5a)} \qquad & x_1x_2 + x_1^2x_2^2 + x_3 \\
\text{(5b)} \qquad & x_1x_3 + x_1^2x_3^2 + x_2 \\
\text{(5c)} \qquad & x_1x_2x_3
\end{align*}
$$

typed as

```
\begin{subequations} \label{E:gp}
   \begin{gather}
      x_{1} x_{2} + x_{1}^{2} x_{2}^{2} + x_{3} \label{E:gp1}\\
      x_{1} x_{3} + x_{1}^{2} x_{3}^{2} + x_{2} \label{E:gp2}\\
      x_{1} x_{2} x_{3} \label{E:gp3}
   \end{gather}
\end{subequations}
```

where `\eqref{E:gp}` prints the group number (5), while

`\eqref{E:gp1}`, `\eqref{E:gp2}`, and `\eqref{E:gp3}`

print the formula numbers (5a), (5b), and (5c), respectively.

5.4 *Aligned columns*

The lines of many multiline formulas are naturally divided into *columns*. In this section, you'll find how to display such formulas with *aligned columns*. All these constructs are implemented with the `align` math environment and its variants.

In section 1.4.2, you saw two simple, one-column examples of aligned columns (called *simple align*) and a special case of aligned columns (called *annotated align*).

The `align` environment creates *multiple* aligned columns; the number of columns is restricted only by the width of the page. In the following example, there are two aligned columns:

$$
\begin{aligned}
\text{(6)} \qquad f(x) &= x + yz & g(x) &= x + y + z \\
h(x) &= xy + xz + yz & k(x) &= (x+y)(x+z)(y+z)
\end{aligned}
$$

typed as

```
\begin{align} \label{E:mm3}
   f(x) &= x + yz       & g(x) &= x + y + z        \\
```

```
h(x) &= xy + xz + yz   & k(x) &= (x + y)(x + z)(y + z)
   \notag
\end{align}
```

In a multicolumn `align` environment, the ampersand (&) doubles as a mark for the *alignment point* and as a *column separator*. In the first line of this formula:

```
f(x) &= x + yz          & g(x) &= x + y + z
```

the two columns are

```
f(x) &= x + yz
```

and

```
g(x) &= x + y + z
```

In each column, you find a single ampersand to mark the alignment point. Of the three & symbols,

- the first & marks the *alignment point* of the first column;
- the second & is a *column separator*; it separates the first and second columns.
- the third & marks the *alignment point* of the second column.

I use the convention that I put a blank on the left of the alignment point & and no space to the right. I put spaces on both sides of & as a column separator.

If the number of columns is three, then there should be five &s in each line. The two even-numbered &s are column separators and the three odd-numbered &s are alignment marks.

Rule ■ If there are n aligned columns, then each line should have $2n - 1$ ampersands; the even numbered &s are column separators, and the odd numbered &s mark the alignment points.

5.4.1   *The subformula rule revisited*

Suppose that you want to align the formula

$$x_1 + y_1 + \left(\sum_{i<5} \binom{5}{i} + a^2 \right)^2$$

with

$$y + (31 + a^2)^2$$

so that the $+ a^2$ in the first formula aligns with the $+ a^2$ in the second formula. Try typing

```
\begin{align}
   x_{1} + y_{1} + \left( \sum_{i < 5} \binom{5}{i}
           &+ a^{2} \right)^{2}\\
   y + (31 &+ a^{2})^{2}
\end{align}
```

When typesetting this formula, you'll get the error message:

```
! Extra }, or forgotten \right.
<template> }
              $}\ifmeasuring@ \savefieldlength@ \fi \set@field ...
...
l.12 \end{align}
```

This alignment structure violated the subformula rule since, for instance,

```
   x_{1} + y_{1} + \left( \sum_{i < 5} \binom{5}{i}
```

is a subformula that LaTeX can't typeset.

As another simple example, try to align the + in $\binom{a+b}{2}$ with the + in $x + y$:

```
\begin{align}
  \binom{a &+ b}{2}}\\
    x &+ y
\end{align}
```

When typesetting this formula, you get the error message:

```
! Argument of \align has an extra }.
<inserted text>
                \par
...
l.9    \binom{a &+ b}{2}}
                        \\
```

Again, LaTeX can't typeset the subformula \binom{a.

Problems of this type can sometimes be overcome with the \phantom command (see section 2.8.1), which can also be used in formulas.

5.4.2 *Align variants*

The align environment has two variants. The first is the flush variant: flalign, which displays the left-most column flush left and the right-most column flush right. Here is formula (6) again followed by the flush variant:

(6) $f(x) = x + yz$ $g(x) = x + y + z$

$h(x) = xy + xz + yz$ $k(x) = (x + y)(x + z)(y + z)$

$$(7) \quad f(x) = x + yz \qquad\qquad\qquad g(x) = x + y + z$$
$$h(x) = xy + xz + yz \qquad k(x) = (x + y)(x + z)(y + z)$$

The variant is typed as follows:

```
\begin{flalign} \label{E:mm3fl}
    f(x) &= x + yz        & g(x) &= x + y + z              \\
    h(x) &= xy + xz + yz & k(x) &= (x + y)(x + z)(y + z)
      \notag
\end{flalign}
```

The second variant is the `alignat` environment. While the `align` environment decides how much space to put between the columns, the `alignat` environment inserts no space between the columns so the user has better control of the spacing. It is important to note that the `alignat` environment has an argument, the number of columns.

Here is formula (6) typed with the `alignat` environment:

```
\begin{alignat}{2} \label{E:mm3A}
    f(x) &= x + yz        & g(x) &= x + y + z              \\
    h(x) &= xy + xz + yz & k(x) &= (x + y)(x + z)(y + z)
      \notag
\end{alignat}
```

and typeset:

$$(8) \qquad\qquad f(x) = x + yz \qquad g(x) = x + y + z$$
$$h(x) = xy + xz + yzk(x) = (x + y)(x + z)(y + z)$$

This did not work very well: `alignat` did not separate the two formulas in the second line. So you must provide the intercolumn spacing. For instance, if you want \qquad space between the columns:

$$(9) \qquad\qquad f(x) = x + yz \qquad\qquad g(x) = x + y + z$$
$$h(x) = xy + xz + yz \qquad k(x) = (x + y)(x + z)(y + z)$$

type it as

```
\begin{alignat}{2} \label{E:mm3B}
    f(x) &= x + yz        & \qquad g(x) &= x + y + z              \\
    h(x) &= xy + xz + yz & \qquad k(x) &= (x + y)(x + z)(y + z)
      \notag
\end{alignat}
```

The `alignat` environment is especially appropriate with annotated align; one would normally put \quad between the formula and the text. To achieve this effect:

$$(10) \qquad \begin{aligned} x &= x \wedge (y \vee z) && \text{(by distributivity)} \\ &= (x \wedge y) \vee (x \wedge z) && \text{(by condition (M))} \\ &= y \vee z \end{aligned}$$

type

```
\begin{alignat}{2} \label{E:mm4}
    x &= x \wedge (y \vee z) & &\quad\text{(by distributivity)}\\
        &= (x \wedge y) \vee (x \wedge z) & &
        \quad\text{(by condition (M))}\notag\\
        &= y \vee z \notag
\end{alignat}
```

For historical reasons, it's appropriate to mention the eqnarray math environment of LaTeX, the ancestor of `align`. Here is an example:

```
\begin{eqnarray}
    x & = & 17y \\
    y & > & a + b + c
\end{eqnarray}
```

which prints

$$(11) \qquad\qquad x \;=\; 17y$$

$$(12) \qquad\qquad y \;>\; a+b+c$$

You can type the same formula with `align`:

```
\begin{align}
    x   & =   17y \\
    y   & >   a + b + c
\end{align}
```

which prints

$$(13) \qquad\qquad x = 17y$$

$$(14) \qquad\qquad y > a+b+c$$

As you can see, with eqnarray, the spaces around = and > come out too wide by default; the spacing in an eqnarray environment is based on the spacing of the columns (which you can change).

5.4.3 *Intertext*

The \intertext command places a line (or more) of text in the middle of an aligned environment (align or variants). For instance, to obtain the following:

$$(15) \qquad h(x) = \int \left(\frac{f(x) + g(x)}{1 + f^2(x)} + \frac{1 + f(x)g(x)}{\sqrt{1 - \sin x}} \right) dx$$

The reader may find the following form easier to read:

$$= \int \frac{1 + f(x)}{1 + g(x)} \, dx - 2 \arctan(x - 2)$$

type:

```
\begin{align} \label{E:mm10}
    h(x) &= \int \left(
                    \frac{ f(x) + g(x) }
                         {1 + f^{2}(x)} +
                    \frac{1 + f(x)g(x)}
                         { \sqrt{1 - \sin x} }
                \right) \, dx\\
    \intertext{The reader may find the following form easier to
    read:}
       &= \int \frac{1 + f(x)}
                    {1 + g(x)}
           \, dx - 2 \arctan(x - 2) \notag
\end{align}
```

Observe how the equal sign in the first formula is aligned with the equal sign in the second formula, even though a line of text separates the two.

Here is another example with align*:

$$f(x) = x + yz \qquad\qquad\qquad g(x) = x + y + z$$

The reader also may find the following polynomials useful:

$$h(x) = xy + xz + yz \qquad\qquad k(x) = (x + y)(x + z)(y + z)$$

typed as

```
\begin{align*}
    f(x) &= x + yz & \qquad g(x) &= x + y + z \\
    \intertext{The reader also may find the following
    polynomials useful:}
    h(x) &= xy + xz + yz
                    & \qquad k(x) &= (x + y)(x + z)(y + z)
\end{align*}
```

The \intertext command must follow one of the line separators \\ or *
(see section 5.8). If you violate this rule, you get the error message:

```
! Misplaced \noalign.
\intertext #1->\noalign
                        {\penalty \postdisplaypenalty \vskip ...
l.11 \end{align*}
```

The text in \intertext can be centered using a center environment or with
the \centering command declaration (see section 3.8).

5.5 *Aligned subsidiary math environments*

A *subsidiary math environment* is a math environment that must appear inside another math environment; think of it as a "large" math symbol you create.

In this section, we discuss aligned subsidiary math environments. Adjusted
subsidiary math environments (including cases) will come up in section 5.6.

5.5.1 *Subsidiary variants of aligned math environments*

The align (see section 5.4) and the gather (see section 5.1) environments have
subsidiary versions: aligned and gathered, respectively.

To obtain this display:

$$
\begin{aligned}
x &= 3 + a + \alpha \\
y &= 4 + b \\
z &= 5 + c \\
u &= 6 + d
\end{aligned}
\qquad \text{or} \qquad
\begin{gathered}
x = 5 + a + \alpha \\
y = 12 \\
z = 13 \\
u = 11 + d
\end{gathered}
$$

type:

```
\[
   \begin{aligned}
      x &= 3 + a + \alpha \\
      y &= 4 + b\\
      z &= 5 + c \\
      u &=6 + d
   \end{aligned}
   \text{\qquad or\qquad}
   \begin{gathered}
      x = 5 + a + \alpha \\
      y = 12 \\
      z = 13 \\
      u = 11 + d
   \end{gathered}
\]
```

Note how this list of aligned formulas

$$
\begin{aligned}
x &= 3 + a + \alpha \\
y &= 4 + b \\
z &= 5 + c \\
u &= 6 + d
\end{aligned}
$$

and this list of centered formulas

$$
\begin{gathered}
x = 5 + a + \alpha \\
y = 12 \\
z = 13 \\
u = 11 + d
\end{gathered}
$$

are treated as individual "large symbols".

The aligned and gathered subsidiary math environments follow the same rules as align and gather, respectively. aligned allows any number of columns, but the user has to specify the intercolumn spacing as in alignat.

You can use the aligned subsidiary math environment to rewrite formula (5) from section 1.4.2 so that the formula number is centered between the two lines:

$$
\begin{aligned}
h(x) &= \int \left(\frac{f(x) + g(x)}{1 + f^2(x)} + \frac{1 + f(x)g(x)}{\sqrt{1 - \sin x}} \right) dx \\
&= \int \frac{1 + f(x)}{1 + g(x)} \, dx - 2 \arctan(x - 2)
\end{aligned}
\tag{16}
$$

This is typed as

```
\begin{equation} \label{E:mm5}
   \begin{aligned}
      h(x) &= \int \left(
                        \frac{ f(x) + g(x) }
                             { 1 + f^{2}(x) } +
                        \frac{ 1 + f(x)g(x) }
                             { \sqrt{1 - \sin x} }
                 \right) \, dx\\
          &= \int \frac{ 1 + f(x) }
                       { 1 + g(x) } \, dx - 2 \arctan (x - 2)
   \end{aligned}
\end{equation}
```

See section 5.5.2 for a better way to split a long formula.

Symbols, as a rule, are "centrally aligned". This is not normally an issue with math symbols, but it may be important with the large symbols created by subsidiary math environments. Some subsidiary math environments (aligned, gathered, and array) take c, t, or b as an optional argument to force centered, top, or bottom alignment, respectively. The default is c (centered). To obtain

$$
\begin{aligned}
x &= 3 + a + \alpha \\
y &= 4 + b \\
z &= 5 + c \\
u &= 6 + d
\end{aligned}
\qquad \text{or} \qquad
\begin{gathered}
x = 5 + a + \alpha \\
y = 12 \\
z = 13 \\
u = 11 + d
\end{gathered}
$$

for example, type

```
\[
   \begin{aligned}[b]
      x &= 3 + a + \alpha \\
      y &= 4 + b\\
      z &= 5 + c \\
      u &=6 + d
   \end{aligned}
   \text{\qquad or\qquad}
   \begin{gathered}[b]
      x = 5 + a + \alpha \\
      y = 12 \\
      z = 13 \\
      u = 11 + d
   \end{gathered}
\]
```

There is no numbering or \tag-ing in subsidiary environments; LaTeX does not number or tag a "symbol".

5.5.2   *Split*

The split *subsidiary math environment* is used to split a (very long) formula into aligned parts. The math environment containing it considers split a single equation, so it generates only one number for the formula. Example:

$$
\begin{split}
(x_1 x_2 x_3 x_4 x_5 x_6)^2 & \\
+ (x_1 x_2 x_3 x_4 x_5 + x_1 x_3 x_4 x_5 x_6 + x_1 x_2 x_4 x_5 x_6 + x_1 x_2 x_3 x_5 x_6)^2 & \\
+ (x_1 x_2 x_3 x_4 + x_1 x_2 x_3 x_5 + x_1 x_2 x_4 x_5 + x_1 x_3 x_4 x_5)^2 &
\end{split}
\tag{17}
$$

typed as

```
\begin{equation} \label{E:mm6}
   \begin{split}
        &(x_{1} x_{2} x_{3} x_{4} x_{5} x_{6})^{2}\\
        &+ (x_{1} x_{2} x_{3} x_{4} x_{5}
```

```
    + x_{1} x_{3} x_{4} x_{5} x_{6}
    + x_{1} x_{2} x_{4} x_{5} x_{6}
    + x_{1} x_{2} x_{3} x_{5} x_{6})^{2}\\
   &+ (x_{1} x_{2} x_{3} x_{4} + x_{1} x_{2} x_{3} x_{5}
    + x_{1} x_{2} x_{4} x_{5} + x_{1} x_{3} x_{4} x_{5})^{2}
 \end{split}
\end{equation}
```

Rule ■ split subsidiary math environment

1. split must be inside another environment:

   equation, align, gather, flalign, gathered

   or their *-ed variants.
2. A split formula has only one number or tag—automatically generated or declared by \tag, or no tag if so declared by \notag.
3. \tag and \notag must precede \begin{split} or follow \end{split}.

If you use split outside a displayed math environment, you get the error message:

```
! Package amsmath Error: \begin{split} won't work here.
  ...
```

1.7   \begin{split}

You can put align* and split in gather:

```
\begin{gather} \label{E:mm7}
  \begin{split}
    f &= (x_{1} x_{2} x_{3} x_{4} x_{5} x_{6})^{2}\\
      &= (x_{1} x_{2} x_{3} x_{4} x_{5}
       + x_{1} x_{3} x_{4} x_{5} x_{6}
       + x_{1} x_{2} x_{4} x_{5} x_{6}
       + x_{1} x_{2} x_{3} x_{5} x_{6})^{2}\\
      &= (x_{1} x_{2} x_{3} x_{4}
       + x_{1} x_{2} x_{3} x_{5}
       + x_{1} x_{2} x_{4} x_{5}
       + x_{1} x_{3} x_{4} x_{5})^{2}
  \end{split}\\
  \begin{align*}
    g &= y_{1} y_{2} y_{3}\\
```

```
        h &= z_{1}^{2} z_{2}^{2} z_{3}^{2} z_{4}^{2}
  \end{align*}
\end{gather}
```

which prints:

$$f = (x_1 x_2 x_3 x_4 x_5 x_6)^2$$

(18)
$$= (x_1 x_2 x_3 x_4 x_5 + x_1 x_3 x_4 x_5 x_6 + x_1 x_2 x_4 x_5 x_6 + x_1 x_2 x_3 x_5 x_6)^2$$

$$= (x_1 x_2 x_3 x_4 + x_1 x_2 x_3 x_5 + x_1 x_2 x_4 x_5 + x_1 x_3 x_4 x_5)^2$$

$$g = y_1 y_2 y_3$$
$$h = z_1^2 z_2^2 z_3^2 z_4^2$$

Here is an example to illustrate split in align:

$$f = (x_1 x_2 x_3 x_4 x_5 x_6)^2$$

(19)
$$= (x_1 x_2 x_3 x_4 x_5 + x_1 x_3 x_4 x_5 x_6 + x_1 x_2 x_4 x_5 x_6 + x_1 x_2 x_3 x_5 x_6)^2$$

$$= (x_1 x_2 x_3 x_4 + x_1 x_2 x_3 x_5 + x_1 x_2 x_4 x_5 + x_1 x_3 x_4 x_5)^2,$$

(20) $$g = y_1 y_2 y_3.$$

which is typed as follows

```
\begin{align} \label{E:mm8}
  \begin{split}
    f &= (x_{1} x_{2} x_{3} x_{4} x_{5} x_{6})^{2}\\
      &= (x_{1} x_{2} x_{3} x_{4} x_{5}
       + x_{1} x_{3} x_{4} x_{5} x_{6}
       + x_{1} x_{2} x_{4} x_{5} x_{6}
       + x_{1} x_{2} x_{3} x_{5} x_{6})^{2}\\
      &= (x_{1} x_{2} x_{3} x_{4}
       + x_{1} x_{2} x_{3} x_{5}
       + x_{1} x_{2} x_{4} x_{5}
       + x_{1} x_{3} x_{4} x_{5})^{2},
  \end{split}\\
    g &= y_{1} y_{2} y_{3}. \label{E:mm9}
\end{align}
```

Notice the \\ command following \end{split} to separate the lines for align.

The amsmath package is very careful where it places the formula number. If amsmath decides to shift the number, you can further adjust it with the \raisetag command; for instance,

```
\raisetag{6pt}
```

will shift the equation number upwards by 6 points. Note that if the formula number is not shifted by amsmath, then the command has no effect.

See the $\mathcal{A}_{\mathcal{M}}\mathcal{S}$-LaTeX document class options in section 8.4 and the amsmath package options in section 8.4.1 for options that modify the placement of equation numbers.

5.6   *Adjusted columns*

In an *adjusted* multiline math environment, the columns are adjusted: *centered*, or *flush left*, or *flush right*, instead of aligned (as in section 5.4).

In sections 5.1 and 5.2, we discussed two adjusted (one-column) math environments, gather and multline. All the other adjusted constructs are *subsidiary* math environments. For example, matrix is multicolumn, centered:

$$\begin{pmatrix} a+b+c & uv & x-y & 27 \\ a+b & u+v & z & 1340 \end{pmatrix} = \begin{pmatrix} 1 & 100 & 115 & 27 \\ 201 & 0 & 1 & 1340 \end{pmatrix}$$

and array is multicolumn adjusted (centered, or flush left, or flush right):

$$\left(\begin{array}{cccc} a+b+c & uv & x-y & 27 \\ a+b & u+v & z & 1340 \end{array} \right) = \left(\begin{array}{cccc} 1 & 100 & 115 & 27 \\ 201 & 0 & 1 & 1340 \end{array} \right)$$

(In this example, there are three centered columns and one flush right column.) A variant, cases, has two columns flush left:

$$(21) \qquad f(x) = \begin{cases} -x^2, & \text{if } x < 0; \\ \alpha + x, & \text{if } 0 \le x \le 1; \\ x^2, & \text{otherwise.} \end{cases}$$

5.6.1   *Matrices*

Use the matrix subsidiary math environment to typeset matrices. Here is an example:

```
\begin{equation*}
  \left(
  \begin{matrix}
     a + b + c & uv    & x - y & 27   \\
     a + b     & u + v & z     & 1340
  \end{matrix}
  \right) =
  \left(
  \begin{matrix}
     1   & 100 & 115 & 27   \\
     201 & 0   & 1   & 1340
```

```
    \end{matrix}
    \right)
\end{equation*}
```

which prints:

$$\begin{pmatrix} a+b+c & uv & x-y & 27 \\ a+b & u+v & z & 1340 \end{pmatrix} = \begin{pmatrix} 1 & 100 & 115 & 27 \\ 201 & 0 & 1 & 1340 \end{pmatrix}$$

If you use `matrix` on its own:

```
\begin{matrix}
  a + b + c & uv    & x - y & 27 \\
  a + b     & u + v & z     & 134
\end{matrix}
```

you get the error message:

```
! Missing $ inserted.
<inserted text>
                $
1.5 \begin{matrix}
```

reminding you that `matrix` is a *subsidiary* math environment.

The amsmath package provides a matrix of up to 10 centered columns. If you need more columns, you have to ask for them. The following example sets the number of columns to 12:

```
\begin{equation} \label{E:mm12}
  \setcounter{MaxMatrixCols}{12}
  \begin{matrix}
    1 & 2 & 3 & 4 & 5 & 6 & 7 & 8 & 9 & 10 & 11 & 12\\
    1 & 2 & 3 & \hdotsfor{7}              & 11 & 12
  \end{matrix}
\end{equation}
```

which prints:

$$(22) \qquad \begin{matrix} 1 & 2 & 3 & 4 & 5 & 6 & 7 & 8 & 9 & 10 & 11 & 12 \\ 1 & 2 & 3 & \multicolumn{7}{c}{\dotfill} & 11 & 12 \end{matrix}$$

We discuss `\setcounter` and counters further in section 9.3.1.

You can have dots span a number of columns with the `\hdotsfor` command; the argument specifies the number of columns (which is one more than the number of `&`s the command replaces). The `\hdotsfor` command must be at the beginning of a row or it must follow an `&`. If you violate this rule, you get the error message:

```
! Misplaced \omit.
\multispan #1->\omit
                        \mscount #1\relax \loop \ifnum \mscount ...

1.12 \end{equation}
```

The \hdotsfor command has an optional argument, a number, which multiplies the spacing between the dots; the default is 1.

Matrix variants

Using delimiters (see section 4.6.1), a matrix may be enclosed in a number of ways:

$$
\begin{matrix} a+b+c & uv \\ a+b & c+d \end{matrix}
\qquad
\begin{pmatrix} a+b+c & uv \\ a+b & c+d \end{pmatrix}
\qquad
\begin{bmatrix} a+b+c & uv \\ a+b & c+d \end{bmatrix}
$$

$$
\begin{vmatrix} a+b+c & uv \\ a+b & c+d \end{vmatrix}
\qquad
\begin{Vmatrix} a+b+c & uv \\ a+b & c+d \end{Vmatrix}
\qquad
\left(\begin{matrix} a+b+c & uv \\ a+b & c+d \end{matrix} \right]
$$

The first matrix is typed as follows:

```
\begin{matrix}
   a + b + c & uv    \\
   a + b     & c + d
\end{matrix}
```

and the next four with the pmatrix, bmatrix, vmatrix, and Vmatrix environments, respectively. The last one is typed:

```
\left(
\begin{matrix}
   a + b + c & uv    \\
   a + b     & c + d
\end{matrix}
\right]
```

Here is another example, utilizing \vdots and \ddots:

```
\begin{equation*}
   \begin{pmatrix}
       1     & 0     & \dots & 0      \\
       0     & 1     & \dots & 0      \\
       \vdots & \vdots & \ddots & \vdots\\
       0     & 0     & \dots & 1
   \end{pmatrix}
\end{equation*}
```

which prints:

$$\begin{pmatrix} 1 & 0 & \cdots & 0 \\ 0 & 1 & \cdots & 0 \\ \vdots & \vdots & \ddots & \vdots \\ 0 & 0 & \cdots & 1 \end{pmatrix}$$

Small matrix

If you put a matrix in an inline math formula, it may be too large; instead, use the smallmatrix environment. Compare $\begin{pmatrix} a+b+c & uv \\ a+b & c+d \end{pmatrix}$, typed as

```
$\begin{pmatrix}
   a + b + c & uv   \\
   a + b     & c + d
\end{pmatrix}$
```

with the small matrix in this line $\left(\begin{smallmatrix} a+b+c & uv \\ a+b & c+d \end{smallmatrix} \right)$ typed as

```
$\left(
\begin{smallmatrix}
   a + b + c & uv   \\
   a + b     & c + d
\end{smallmatrix}
\right)$
```

There are no variants of smallmatrix similar to the variants of matrix; use \left and \right with delimiters to enclose a small matrix. Also, the \hdotsfor command does not work in a small matrix

5.6.2 *Arrays*

The matrix subsidiary math environment of the amsmath package and the array subsidiary math environment of LATEX are similar.

The first array in the introduction to section 5.6 is typed as follows:

```
\begin{equation*}
   \left(
   \begin{array}{cccc}
      a + b + c & uv    & x - y & 27 \\
      a + b     & u + v & z     & 134
   \end{array}
   \right)
\end{equation*}
```

which prints:

$$\left(\begin{array}{cccc} a+b+c & uv & x-y & 27 \\ a+b & u+v & z & 134 \end{array}\right)$$

Rule ■ array subsidiary math environment

1. The columns are separated by &.
2. The argument of \begin{array} is mandatory: it's a string made up of the letters l, r, or c, one letter for each column; a column is flush left, centered, or flush right, if the corresponding letter is l, c, or r, respectively.

The following is an example of an array that could not have been made with matrix:

$$\begin{array}{cccc} a+b+c & uv & x-y & 27 \\ a+b & u+v & z & 134 \end{array}$$

since the last column is flush right. (Of course, this is not quite true: in a matrix environment, \hfill 27 will flush the entry 27 right—see section 2.8.4.)

If the argument of \begin{array} is missing:

```
\begin{equation}
   \begin{array}
      a + b + c & uv       & x - y & 27 \\
      a + b     & u + v    & z     & 134
   \end{array}
\end{equation}
```

you get the error message:

```
! LaTeX Error: Illegal character in array arg.
```

```
1.7 a
       + b + c & uv        & x - y & 27 \\
```

If the amsmath package is used, the error message is

```
! Extra alignment tab has been changed to \cr.
<recently read> \endtemplate
```

```
1.14  \end{equation}
```

which is almost the same as the LATEX message if you change the first formula in the array to c + b + a. (Note that the first character in c + b + a is **not** an "Illegal character in array arg.")

If the closing brace of the argument of \begin{array} is missing:

```
\begin{equation}
  \begin{array}{cccc
      a + b + c & uv     & x - y & 27 \\
      a + b     & u + v & z     & 134
  \end{array}
\end{equation}
```

you get the error message:

```
Runaway argument?
{cccc a + b + c & uv     & x - y & 27 \\ a + b      & u + v \ETC.
! Paragraph ended before \@array was complete.
<to be read again>
                \par
```

In fact, the argument of array can take more than stated in the rule; it can take any argument the tabular environment can take (see section 3.7).

5.6.3   *Cases*

The cases environment is also a subsidiary math environment. Here is the example from the introduction to this section:

$$
f(x) = \begin{cases}
-x^2, & \text{if } x < 0; \\
\alpha + x, & \text{if } 0 \leq x \leq 1; \\
x^2, & \text{otherwise.}
\end{cases}
$$

It is typed as

```
\begin{equation}
  f(x)=
  \begin{cases}
    -x^{2},        &\text{if $x < 0$;}              \\
    \alpha + x,    &\text{if $0 \leq x \leq 1$;}\\
    x^{2},         &\text{otherwise.}
  \end{cases}
\end{equation}
```

It would be easy to code the cases environment as a special case of the array environment.

5.7 *Commutative diagrams*

The amscd package provides the CD subsidiary math environment to typeset simple commutative diagrams. For instance, to obtain

type

```
\[
    \begin{CD}
        A           @>>>        B   \\
        @VVV                    @VVV\\
        C           @=          D
    \end{CD}
\]
```

A commutative diagram is a matrix made up of two kinds of rows: *horizontal rows*, that is, rows with horizontal arrows; and *vertical rows*, that is, rows with vertical arrows. For example,

```
A           @>>>        B
```

is a typical horizontal row. It defines two columns and a connecting horizontal arrow @>>>. There may be more than two columns, as in:

```
A   @>>>    B   @>>>    C   @=   D   @<<<   E   @<<<   F
```

The connecting piece can be an extendible arrow @>>> or @<<<, or it could be @=, an extendible equal sign. The label on top of an extendible arrow should be typed between the first and second > (or <) symbols, while the label underneath should be typed between the second and third > (or <) symbols.

The following is a typical vertical row containing vertical arrows:

```
@VVV        @VVV        @AAA
```

@VVV is a down arrow and @AAA is an up arrow. The @V{*label*}VV command puts *label* to the left while @VV{*label*}V puts *label* to the right; similar rules apply to @AAA. The vertical arrows are placed in the columns from the first on.

These constructs are also illustrated in

$$\begin{CD} \mathbb{C} @>H_1>> \mathbb{C} @>H_2>> \mathbb{C} \\ @VP_{c,3}VV @VP_{\bar{c},3}VV @VVP_{-c,3}V \\ \mathbb{C} @>H_1>> \mathbb{C} @>H_2>> \mathbb{C} \end{CD}$$

typed as

```
\[
   \begin{CD}
      \mathbb{C} @>H_{1}>> \mathbb{C} @>H_{2}>> \mathbb{C}   \\
      @VP_{c,3}VV   @VP_{\bar{c},3}VV   @VVP_{-c,3}V   \\
      \mathbb{C} @>H_{1}>> \mathbb{C} @>H_{2}>> \mathbb{C}
   \end{CD}
\]
```

Here is a more complicated example, followed by its source, utilizing also the \text command from the amstext package:

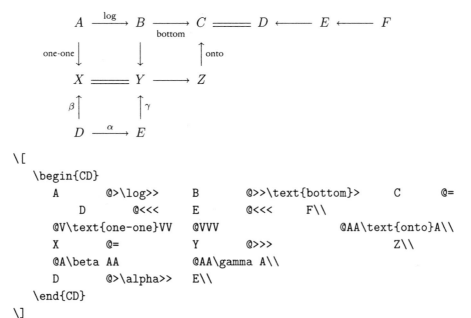

```
\[
   \begin{CD}
      A           @>\log>>       B            @>>\text{bottom}>      C        @=
           D           @<<<      E            @<<<       F\\
      @V\text{one-one}VV         @VVV                               @AA\text{onto}A\\
      X           @=            Y            @>>>                    Z\\
      @A\beta AA                 @AA\gamma A\\
      D           @>\alpha>>     E\\
   \end{CD}
\]
```

More complicated diagrams should be done with a drawing (or drafting) program or with specialized packages; see the CTAN subdirectories

```
tex-archive/macros/generic/diagrams
tex-archive/macros/latex/supported
```

for some diagram packages and Appendix G on how to get them.

5.8   Pagebreak

By default, the math environments of this chapter do not allow pagebreaks. For instance, a pagebreak in cases is obviously not desirable, but it may be permissible in align or gather. You have to decide whether to allow pagebreaks. To allow pagebreaks, use the

```
\allowdisplaybreaks
```

command. It'll allow pagebreaks in a multiline math environment within its scope. For instance,

```
{\allowdisplaybreaks
\begin{align} \label{E:mm13}
   a &= b + c,\\
   d &= e + f,\\
   x &= y + z,\\
   u &= v + w.
\end{align}
}% end of \allowdisplaybreaks
```

allows a pagebreak after any one of the first three lines.

Within the scope of \allowdisplaybreaks, * prohibits a break after that line. Also, the line separators \\ and * can be modified to add some interline space as in section 2.7.2.

Just before the line separator \\, you can put \displaybreak to force a break, or

```
\displaybreak[0]
```

to allow one. \displaybreak[n], where n is 1, 2, or 3, give the intermediate steps between allowing and forcing. \displaybreak[4] is the same as \displaybreak.

preamble

top matter

section

section

section

bibliography

index

top matter

main matter

bottom matter

PART III

Document structure

CHAPTER 6

LATEX documents

In this chapter, we take up the organization of a document. Section 6.1 discusses the document structure in general, and section 6.2 presents the preamble. Section 6.3 discusses the front matter, including the abstract environment and the table of contents. Section 6.4 presents the main matter, including sectioning, cross-referencing, tables, and figures. Section 6.5 covers the back matter, including the bibliography and the index.

These sections discuss the logical design of a LATEX document. The visual design is left to a large extent to the document class. In the last section, we briefly discuss an aspect of visual design, the page style.

For an article, you may safely ignore much of what is discussed here. Start your article with a template that includes a preamble and top matter, and defines the proclamations. You created such a template for the article document class in section 1.6; this is done for the more detailed amsart document class in section 8.3.

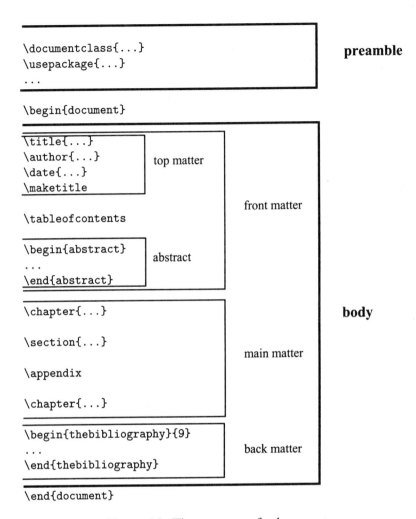

Figure 6.1: The structure of a document

6.1   *The structure of a document*

The source file of a LᴬTEX document is divided into two parts, preamble and body (see Figure 6.1).

Preamble is the portion of the source file before the

```
\begin{document}
```

command; it contains definitions and instructions that affect the entire document.

Body is the document environment itself. It contains everything to be printed.

These statements oversimplify the situation somewhat. For instance, you can define a command in the preamble that typesets some text; the command is used in the body, but what is actually typeset is in the preamble. Nevertheless, I hope the division between the preamble and the body is clear.

The body is further divided into three parts:

Front matter includes, for the most part, the material that goes to the front of the document. For a typical document, this includes the *top matter* from which the title page is constructed and an optional abstract. For longer documents, the front matter may include a table of contents, dedication, preface, and so on.

Main matter is the main part of the document, including any optional appendices.

Back matter consists of material that goes to the back of the document. For a typical document, this consists primarily of the bibliography. For longer documents, the back matter includes the index, and various other matters, such as the colophon, afterword, and so on.

6.2   *The preamble*

You were introduced to the preamble of a document in section 1.5. As you recall, the preamble contains the crucial \documentclass line, naming the document class and the options that modify its behavior. For instance,

```
\documentclass[draft,twocolumn]{article}
```

names the document class article, with options draft, which paints a slug on the margin indicating lines that are too wide (see section 2.7.1), and twocolumn, which prints the document in two-column format (see section 7.1.2).

The \documentclass command is usually followed by the \usepackage commands, which invoke LaTeX enhancements, called *packages*. For instance,

```
\usepackage{amssymb}
```

invokes an $\mathcal{AMS}$ package that defines the symbol names (see section 8.5), whereas

```
\usepackage[reqno]{amsmath}
```

invokes the amsmath package with the reqno option to place equation numbers on the right (see section 8.4). Document class options are passed on to the packages as possible options, so

```
\documentclass[reqno]{amsart}
```

also invokes the amsmath package with the reqno option.

Document class files are designated with the cls extension, while package files are denoted by the sty extension; so the document class article is defined in the article.cls file, while the amsmath package is defined in the amsmath.sty file. You may define your own packages, such as the lattice package in section 9.5.

The preamble normally contains the user-defined commands (see Chapter 9) for the document, and the definitions of proclamations (also called theorem-like structures—see section 3.4).

There are a few commands that *must* be placed in the preamble. Two such examples are the \DeclareMathOperator command (see section 4.7.2) and the \numberwithin command (see section 4.3).

There is an important command that may only be placed *before* the

\documentclass

line:

\NeedsTeXFormat{LaTeX2e}[1994/12/01]

This requires that the version of LaTeX issued on December 1, 1994 or later be used. If your document is typeset with LaTeX 2.09, you get an informative error message. Use the \NeedsTeXFormat command with a date if your document contains a feature that was introduced on the date specified or if an earlier version had a bug that would materially affect the typesetting of your document. For instance, if you use the text symbol • (\textbullet), which was introduced on December 1, 1994, then you should have the \NeedsTeXFormat line as shown above.

Finally, there is one environment that may only be in the preamble, preceding the \documentclass command, namely, the filecontents environment (see section 2.11.2).

6.3   Front matter

The front matter of an article, as a rule, contains the top matter from which the title page is put together and, optionally, an abstract.

Discussion of the top matter should take place in the context of a particular document class. This was done for the article document class in section 1.7; for the amsart document class, it'll be done in section 8.2.

Long documents, such as books, have rather complicated front matter; see the discussion in section 7.1. In this section, we'll only discuss the abstract and the table of contents.

6.3.1   Abstract

Most standard document classes (except those for letters and books) provide an abstract. The abstract is typed in an abstract environment.

The document class formats the heading ABSTRACT (or some variant) and typesets the text of the abstract, as a rule, in a smaller type with wider margins.

Tip   Do not insert an `abstract` environment and postpone typing it until later, because the `abstract` environment can't be empty.

If it's empty, you get the error message

```
! LaTeX Error: Something's wrong--perhaps a missing \item.
```

Either comment out the `abstract` environment, or insert something temporary, such as "`Yet to do!`"

The $\mathcal{AMS}$ document classes require that you place the abstract *before* the

```
\maketitle
```

command; see for instance (on page 364), the abstract in the `sampart.tex` sample article. If you forget, you get the warning:

```
Class amsart Warning: Abstract should precede \maketitle in
AMS documentclasses; reported on input line 21.
```

6.3.2   *Table of contents*

LaTeX creates an auxiliary file, called the `toc` file, to be used by the command

```
\tableofcontents
```

the next time the document is typeset. If the source file is `myart.tex`, the `toc` file is called `myart.toc`. This file lists all the sectioning units (parts, chapters, sections, appendices, and so on), as well as their titles and page numbers.

If you already have a `toc` file, `\tableofcontents` instructs LaTeX to create a new `toc` file, and make a table of contents from the previously created `toc` file. The table of contents is inserted at the point where the `\tableofcontents` command appears in the front matter.

You can add a line to the table of contents, formatted like a section title, say, with the command:

```
\addcontentsline{toc}{section}{line_to_be_added}
```

which is placed in the source file as if it were a sectioning command. There are three arguments. The first argument informs LaTeX that a line, the third argument, should be added to the `toc` file. The second argument specifies how the line is to be formatted. In the example, the second argument is `section`, so the line will be formatted as a section title in the table of contents. The second argument must be the name of a sectioning command (part, chapter, section, subsection, subsubsection, paragraph, or subparagraph).

You can add a nonformatted line to the table of contents with the command:

\addtocontents{toc}{*line_to_be_added*}

This can also be used to help format the table of contents. For instance, if you want to add some space before a part, insert the following line before the sectioning command for the part:

\addtocontents{toc}{\protect\vspace{10pt}}

The toc file is easy to read; you'll find that the lines are self-explanatory. These are typical lines (they would look different with an $\mathcal{AMS}$ document class):

```
\contentsline {section}{\numberline {5-4.}Top matter}{119}
\contentsline {subsection}{\numberline {5-4.1.}Article info}{119}
\contentsline {subsection}{\numberline {5-4.2.}Author info}{121}
```

You can influence which levels of sectioning go into the table of contents (see section 9.3.1). Section 2.4 of *The LATEX Companion* lists the style parameters of the table of contents. It also shows how to define new toc-like files and use multiple tables of contents.

Tip   You may have to **typeset the document three times** to get the table of contents and the rest of the document right.

The first typesetting *creates* the toc file. The second typesetting *inserts* the table of contents with the old page numbers, records the correct page numbers and cross-references in the aux file, and generates a new toc file with the correct page numbers. The third typesetting uses these new aux and toc files to typeset the document correctly.

Recall from section 2.10.1 that fragile commands in a movable argument must be \protect-ed. Here is a simple example using the table of contents. If the document contains the following \section command

\section{The function \(f(x^{2}) \)}

then the title is stored in the toc file as

```
\contentsline {section}{\numberline {1}
The function \relax $ f(x^{2})\relax \GenericError { }
. . .
```

! LaTeX Error: Bad math environment delimiter.

. . .

```
1.1 ...continue without it.}}{1}
```

Normally, error messages refer to a line in the source file; in this case, the error message refers to a line in the toc file.

The correct form for this section title is

```
\section{The function \protect\( f(x^{2}) \protect\)}
```

or even simpler

```
\section{The function $f(x^{2})$}
```

6.4 *Main matter*

The main matter contains the most essential parts of the document, including the appendices.

We discuss sectioning in section 6.4.1, cross-referencing in section 6.4.2, and tables and figures in section 6.4.3. Cross-referencing is typically used only in the main matter, although it could also be used elsewhere.

6.4.1 *Sectioning*

The main matter of a typical document is divided into *sections* and of a longer document into *chapters*. There may also be *parts*.

Section

LaTeX is instructed to start a section with the \section command. This command has an argument: the title of the section. The argument of the \section command may also be used for the running head and the table of contents (see section 6.3.2). LaTeX will typeset the section number automatically, followed by the title.

Of course, \section may be followed by \label, so you can refer to the number generated by LaTeX. Example:

```
\section{Introduction}\label{S:intro}
```

The command \ref{S:intro} refers to the number of the section.

The \section* command works similarly, except that the section is not numbered. The standard LaTeX document classes suppress the running head and the table of contents entry for the *-ed sections; the $\mathcal{AMS}$ document classes do not.

The \section command may have an optional argument that is used in place of the section title for the running head and the table of contents (see section 6.3.2). Example:

```
\section[The first product construction]{The first product
  construction in the reduced trunk case}
```

This optional argument is typically a shortened version of the full section title.

Other sectioning commands

A section may be subdivided into *subsections*, and also *subsubsections*, *paragraphs*, and *subparagraphs*. Subsections are numbered within a section (in section 1, they are numbered 1.1, 1.2, and so on). Book document classes, as a rule, do not number subsubsections (and lower).

It is important to understand that the five levels of sectioning are not just five different styles of typesetting; there should be no subsection without a section and no subsubsection without a subsection, and so on.

The document class determines how the titles of sections, subsections, and so on, are displayed, and which of them are numbered. Consider the text:

```
\section{Introduction}\label{S:Intro}
We shall discuss the main contributors of this era.
\subsection{Birkhoff's contributions}\label{SS:contrib}
Of course, Garrett Birkhoff is the first in line.
\subsubsection{The years 1935--1945}\label{SSS:1935}
Going to Oxford was a major step.
\paragraph{The first paper}
What should be the definition of a universal algebra?
\subparagraph{The idea}
One should read Whitehead very carefully.
```

Figure 6.2 shows how this text is typeset in the `article` document class, while Figure 6.3 shows the same section in the `amsart` document class. Note that in both document classes, paragraphs and subparagraphs are unnumbered.

The `book` and `report` (and `amsbook`) document classes have a higher level sectioning unit, the chapter, invoked by the command \chapter. There is also one more sectioning unit, the *part*, but it has no effect on other sectioning commands. The corresponding command is \part. It is placed between chapters in longer documents such as books; see the present book for an example. Both \chapter and \part have an argument, the title of the chapter or part, respectively.

There are also *-ed versions of all these commands that suppress the number. If you * a section (chapter), make sure that all subsections (sections), and so on, are also *-ed.

Section 9.3.1 discusses how to change the formatting of the section numbers, and how to determine which sectioning levels are numbered.

In the main matter, the \appendix command marks the beginning of the appendices. Each subsequent section (or chapter, for the book, report, and `amsbook`

1 Introduction

We shall discuss the main contributors of this era.

1.1 Birkhoff's contributions

Of course, Garrett Birkhoff is the first in line.

1.1.1 The years 1935–1945

Going to Oxford was a major step.

The first paper What should be the definition of a universal algebra?

The idea One should read Whitehead very carefully.

Figure 6.2: Sectioning commands in the `article` document class

1. INTRODUCTION

We shall discuss the main contributors of this era.

1.1. Birkhoff's contributions. Of course, Garrett Birkhoff is the first in line.

1.1.1. *The years 1935–1945.* Going to Oxford was a major step.
The first paper. What should be the definition of a universal algebra?
The idea. One should read Whitehead very carefully.

Figure 6.3: Sectioning commands in the `amsart` document class

document classes) becomes an appendix. For example, in the `article` document class:

```
\appendix
\section{A geometric proof of the Main Theorem}\label{S:geom}
```

produces an appendix with the given title.

Note that appendices may be labeled and cross-referenced. In an appendix (as a section), subsections are numbered A.1, A.2, and so on, while subsubsections in A.1 are numbered A.1.1, A.1.2, and so on; the precise form, of course, depends on the document class.

Sections 2.3.1 and 2.3.2 of *The LATEX Companion* show how to change the layout of section headings.

Equations in sections

Equations (see section 4.3) are numbered sequentially in an article. In the book document classes, however, normally they are numbered from 1 within each chapter. The amsmath package provides the

```
\numberwithin{equation}{section}
```

command to be placed in the preamble. This will cause the equations to be numbered within sections; so in section 1, the equations will be numbered (1.1), (1.2), and so on. If the document class has chapters, you can use the

```
\numberwithin{equation}{chapter}
```

to have the equations numbered within chapters.

6.4.2   *Cross-referencing*

There are three types of cross-referencing in LaTeX:

- Symbolic referencing with \ref
- Page referencing with \pageref
- Bibliographic referencing with \cite

In this section, we discuss the first two; the third is discussed in section 6.5.1 and Chapter 10.

Symbolic referencing

Wherever LaTeX generates a number, you can place a \label command:

```
\label{symbol}
```

Then at any place in the text you can use the \ref command, type:

```
\ref{symbol}
```

to reproduce the number. We call *symbol* the *label*.

You can use labels for sectioning units, equations, figures, tables, items in an enumerated environment (see section 3.1.1), and also for theorems and other proclamations.

If the equation labeled E:int is the fifth in an article, then the label E:int will store the number 5, and \ref{E:int} will produce the number 5. If equations are numbered within sections (see section 6.4.1), say, the equation is the third in section 2, then the label E:int will store the number 2.3 and \ref{E:int} will produce the number 2.3.

Tip **Typeset a document twice** to see a change in a cross-reference.

If you typeset only once, and LaTeX suspects that the cross-references have not been updated, you get a warning:

```
LaTeX Warning: Label(s) may have changed.
Rerun to get cross-references right.
```

Example 1 The title of the present section of this book is typed

```
\section{Main matter}\label{S:MainMatter}
```

So `\ref{S:MainMatter}` will produce the number 6.4.

Example 2

```
\begin{equation}\label{E:int}
    \int_{0}^{\pi} \sin x \, dx = 2.
\end{equation}
```

In this case, `\ref{E:int}` produces the number of the equation. If parentheses are required, you must type

```
(\ref{E:int})
```

or use the `\eqref` command (in the amsmath package), which supplies the parentheses automatically. In fact, `\eqref` does somewhat more: even if the text is emphasized (as in theorems), the parentheses and the number will be upright.

Example 3

```
\begin{theorem}\label{T:fund}
    Statement of theorem.
\end{theorem}
```

The reference `\ref{T:fund}` produces the number of the theorem.

Rule ■ `\label` command
The argument of the `\label` command is a string of letters, punctuation marks, and digits. It is case sensitive, so `S:intro` is different from `S:Intro`.

Tip Place the `\label` command immediately after the command that generates the number.

Tip   Use the nonbreakable space ~ when referencing:

```
see section~\ref{S:Intro}
proved in Theorem~\ref{T:main}
```

It is difficult to overemphasize how useful automatic cross-referencing is in the writing of a document. There are three simple ways to make cross-referencing even more useful.

Tip

1. Utilize user-defined commands to minimize the typing necessary for referencing (see section 9.1.1).
2. Systematize the labels. For example, start the label for a section with `S:`, subsection with `SS:`, subsubsection with `SSS:`, theorem with `T:`, lemma with `L:`, definition with `D:`, and so on.
3. The labels should be meaningful.

While working on an article, typeset it with the labels shown on the margin. Include in the preamble the line:

```
\usepackage{showkeys}
```

which invokes David Carlisle's showkeys package, part of the Tools distribution (see section 7.3.1 and Appendix G). This package shows all symbolic references on the typeset copy. With the `notcite` option (my preference),

```
\usepackage[notcite]{showkeys}
```

showkeys does not show the labels for the bibliographic references. When the document is ready for final typesetting, comment out this line.

Section 2.5 of *The LaTeX Companion* describes a package extending the power of `\ref` and another for referencing other documents.

Closely related to labels are citations (see section 6.5.1).

Page-referencing

The command

```
\pageref{symbol}
```

produces the number of the typeset page corresponding to the location of the command `\label{symbol}`. For example, if the following text is typeset on page 5:

There may be three types of problems with the
construction of such lattices.\label{problem}

and somewhere else you type:

Because of the problems associated with
the construction (see page~\pageref{problem})

LaTeX typesets

Because of the problems associated with the construction (see page 5)

6.4.3 *Tables and figures*

Many documents contain tables and figures. These must be treated in a special way
since they can't be broken across pages. LaTeX provides the following solution. A
table or a figure would normally be moved ("floated") to the top or bottom of the
current page or to the next page (or further).

LaTeX offers two environments to handle tables and figures: the `table` and
the `figure` environments. The two work the same way except that the titles of
the first are collected in a list of tables, while titles of the latter are collected in a
list of figures.

A `table` environment is set up as follows:

```
\begin{table}
    Place the table here
    \caption{name}\label{Ta:xxx}
\end{table}
```

The `table` environment is primarily used for tables made with the `tabular`
environment (see section 3.7). The `\caption` is optional; its argument is typeset
and printed on the page and entered into the list of tables. The optional `\label`
command must be between the caption and `\end{table}`. The label is used to
reference the table number. An optional argument of the `\caption` command can
replace the argument in the list of tables. A table can have more than one caption.

There are many examples of tables in this book; for instance, section 2.4.6 has
three.

Following the `\begin{table}` command, you can specify an optional argument b for bottom of page, h for here, t for top of page, or p for separate page. In
fact, you can list more than one option, for instance,

```
\begin{table}[ht]
```

requests LaTeX to place the table "here" or at the "top" of a page. The default is
[tbp] and the order of the optional arguments is immaterial: [th] is the same as
[ht].

LATEX has more than a dozen parameters that control a complicated algorithm for the placement of a table. If you want to override them *for one table only*, modify your request with !. For instance, [!ht] requests that this table be placed here or at the top even if this placement violates the rules as set by some of the parameters. For a detailed discussion of the floating mechanism, see Chapter 6 of *The LATEX Companion*.

The \suppressfloats command stops LATEX from placing any more tables on the page. There is an optional argument t or b (but not both) prohibiting placement at the top or bottom of the present page. The table that is "suppressed" will go to the next page or further.

The information for the list of tables is placed by LATEX in the lot file if so instructed by the \listoftables command. The list of tables is inserted into the body where the command appears, normally in the front matter.

There are analogs of the table of contents commands discussed in section 6.3.2 for tables. The commands

```
\addcontentsline{lot}{sectioning}{line_to_add}
\addtocontents{lot}{line_to_add}
```

add a line to the list of tables.

Your demands and LATEX's floating mechanism may conflict with one another: LATEX may run out of memory or place the material where you do not want it. Breaking two tables into one sometimes helps. The \clearpage command not only gives a \newpage command, but also instructs LATEX to print out all the tables and figures accumulated, but not yet printed. See also the article section 2.7.3.

Graphics (drawings, scanned images, digitized photos, and so on) can be inserted with a figure environment. Figures have captions that are also numbered by LATEX; if labels are used, these numbers may be cross-referenced. A list of figures (similar to a list of tables) can be compiled with the \listoffigures command. This creates an auxiliary file with the extension lof. The graphic can be made within a picture environment (which is not discussed in this book) or it can be read from a file. The standard way of including a graphics file is with the graphics package by David Carlisle and Sebastian Rahtz, which is a part of the LATEX distribution (see section 7.3).

Using the graphics package, a typical figure is specified as follows:

```
\begin{figure}
   \includegraphics{file}
   \caption{title}\label{Fi:xxx}
\end{figure}
```

This assumes that you have a graphics file in a form that your printer driver can handle, *file* contains the information about its size (encapsulated PostScript pictures do, for example). If you have to scale the graphics image, say to 68% of its original size, use the command

```
\scalebox{.68}{\includegraphics{file}}
```

For instance, the figure on page 51 is included with the commands

```
\begin{figure}[t!]
   \scalebox{.80}{\includegraphics{fig1.art}}
   \caption{The structure of \protect\LaTeX}\label{Fi:StrucLaT}
\end{figure}
```

while the figure on page 53 is included with the commands

```
\begin{figure}[p]
   \scalebox{.65}{\includegraphics{fig2.art}}
   \caption{Using \protect\LaTeX}\label{Fi:UsingLaT}
\end{figure}
```

The `\scalebox` command (provided by the graphics package) can scale text or math as well. For instance,

```
\scalebox{0.5}{\parbox{\textwidth}{
\begin{gather*}
   x_{1} x_{2} + x_{1}^{2} x_{2}^{2} + x_{3}, \\
   x_{1} x_{3} + x_{1}^{2} x_{3}^{2} + x_{2},\\
   x_{1} x_{2} x_{3}.
 \end{gather*}
}}
```

provides a thumbnail sketch of this multiline math formula:
$$x_1 x_2 + x_1^2 x_2^2 + x_3$$
$$x_1 x_3 + x_1^2 x_3^2 + x_2$$
$$x_1 x_2 x_3$$

For more information on graphics, see the *The LATEX Companion*, especially Chapters 10 and 11, and the documentation of the graphics package in the LATEX distribution.

There are also the

```
\addcontentsline{lof}{sectioning}{line_to_add}
\addtocontents{lof}{line_to_add}
```

commands, which add a line to the list of figures.

The above discussion on how to influence LATEX on where to place the tables also applies to figures.

References

[1] Henry H. Albert, *Free torsoids*, Current Trends in Lattice Theory, D. Van Nostrand, 1970.

[2] Henry H. Albert, *Free torsoids*, Current Trends in Lattice Theory (G. H. Birnbaum, ed.), vol. 7, D. Van Nostrand, Princeton-Toronto-London-Melbourne, January 1970, no translation available, pp. 173–215 (German).

[3] Soo-Key Foo, *Lattice Constructions*, Ph.D. thesis, University of Winnebago, 1990.

[4] Soo-Key Foo, *Lattice Constructions*, Ph.D. thesis, University of Winnebago, Winnebago MN, December 1990, final revision not yet available.

[5] Grant H. Foster, *Computational complexity in lattice theory*, tech. report, Carnegie Mellon University, 1986.

[6] Grant H. Foster, *Computational complexity in lattice theory*, Research Note 128A, Carnegie Mellon University, Pittsburgh PA, December 1986, research article in preparation.

[7] Peter Konig, *Composition of functions*, Proceedings of the Conference on Universal Algebra (Kingston, 1969).

[8] Peter Konig, *Composition of functions*, Proceedings of the Conference on Universal Algebra (G. H. Birnbaum, ed.), vol. 7, Canadian Mathematical Society, Queen's Univ., Kingston ON, available from the Montreal office, pp. 1–106 (English).

[9] William A. Landau, *Representations of complete lattices*, Abstract: Notices Amer. Math. Soc., **18**, 937.

[10] William A. Landau, *Representations of complete lattices*, Abstract: Notices Amer. Math. Soc. **18**, 937, December 1975.

[11] George A. Menuhin, *Universal Algebra*, D. van Nostrand, Princeton-Toronto-London-Melbourne, 1968.

[12] George A. Menuhin, *Universal Algebra*, Second ed., University Series in Higher Mathematics, vol. 58, D. van Nostrand, Princeton-Toronto-London-Melbourne, March 1968 (English), no Russian translation.

[13] Ernest T. Moynahan, *On a problem of M. H. Stone*, Acta Math. Acad. Sci. Hungar. **8** (1957), 455–460.

[14] Ernest T. Moynahan, *On a problem of M. H. Stone*, Acta Math. Acad. Sci. Hungar. **8** (1957), 455–460 (English), Russian translation available.

6.5   Back matter

The back matter of an article is very simple, as a rule. It usually contains, as an option, the bibliography. A long document, such as a book, may have a more complicated back matter (see section 7.1). In this section, we only discuss the bibliography.

6.5.1   Bibliography in an article

The simplest way to typeset a bibliography is to type it directly into the article. For an example, see the bibliography in the `intrart.tex` introductory sample article (page 40). A more complete example is shown on the facing page. This sample bibliography contains two each (one short and one long) of the seven most often used kinds of items.

Type the text of the bibliography in a `thebibliography` environment, as in the following examples (you can find these entries in `inbibl.tpl`—see page 4; also, the templates for these entries are reproduced in `article.tpl` following the line `\end{document}`):

```
\begin{thebibliography}{99}
\bibitem{hA70}
   Henry~H. Albert,
   \emph{Free torsoids},
   Current Trends in Lattice Theory, D.~Van Nostrand, 1970.
\bibitem{hA70a}
   Henry~H. Albert,
   \emph{Free torsoids},
   Current Trends in Lattice Theory (G.~H. Birnbaum, ed.), vol.~7,
   D.~Van Nostrand, Princeton-Toronto-London-Melbourne,
   January 1970, no translation available, pp.~173--215 (German).
\bibitem{sF90}
   Soo-Key Foo,
   \emph{Lattice Constructions},
   Ph.D. thesis, University of Winnebago, 1990.
\bibitem{sF90a}
   Soo-Key Foo,
   \emph{Lattice Constructions},
   Ph.D. thesis, University of Winnebago, Winnebago MN,
   December 1990, final revision not yet available.
\bibitem{gF86}
   Grant~H. Foster,
   \emph{Computational complexity in lattice theory},
   tech. report, Carnegie Mellon University, 1986.
```

```
\bibitem{gF86a}
   Grant~H. Foster,
   \emph{Computational complexity in lattice theory},
   Research Note 128A, Carnegie Mellon University,
   Pittsburgh PA, December 1986, research article in preparation.
\bibitem{pK69}
   Peter Konig,
   \emph{Composition of functions},
   Proceedings of the Conference on Universal Algebra
   (Kingston, 1969).
\bibitem{pK69a}
   Peter Konig,
   \emph{Composition of functions},
   Proceedings of the Conference on Universal Algebra
   (G.~H. Birnbaum, ed.),
   vol.~7, Canadian Mathematical Society,
   Queen's Univ., Kingston ON,
   available from the Montreal office, pp.~1--106 (English).
\bibitem{wL75}
   William~A. Landau,
   \emph{Representations of complete lattices},
   Abstract: Notices Amer. Math. Soc., \textbf{18}, 937.
\bibitem{wL75a}
   William~A. Landau,
   \emph{Representations of complete lattices},
   Abstract: Notices Amer. Math. Soc. \textbf{18}, 937,
   December 1975.
\bibitem{gM68}
   George~A. Menuhin,
   \emph{Universal Algebra},
   D.~van Nostrand, Princeton-Toronto-London-Melbourne, 1968.
\bibitem{gM68a}
   George~A. Menuhin,
   \emph{Universal Algebra}, Second ed.,
   University Series in Higher Mathematics, vol.~58,
   D.~van Nostrand, Princeton-Toronto-London-Melbourne,
   March 1968 (English), no Russian translation.
\bibitem{eM57}
   Ernest~T. Moynahan,
   \emph{On a problem of M.~H. Stone},
   Acta Math. Acad. Sci. Hungar. \textbf{8}~(1957), 455--460.
\bibitem{eM57a}
```

```
    Ernest~T. Moynahan,
    \emph{On a problem of M.~H. Stone},
    Acta Math. Acad. Sci. Hungar. \textbf{8}~(1957), 455--460
    (English), Russian translation available.
\end{thebibliography}
```

I use the convention that the label for the \bibitem consists of the initials of the author and the year of publication. A first publication by Andrew B. Reich in 1987 would have the label aR87; the second, aR87a. Of course, you can use any label you choose; however, such conventions help greatly in making the items reusable.

The environment thebibliography has an argument; in the previous example, this argument is 99. This tells LaTeX that the widest reference number it must generate is two characters wide; for fewer than 10 items, use 9 and for 100 or more items use 999.

If the argument of \begin{thebibliography} is missing, you get the error message:

```
! LaTeX Error: Something's wrong--perhaps a missing \item.

1.5 \bibitem
            {hA70}
```

Each bibliographic item is introduced with \bibitem, which is just like the \label command. In the text, use \cite, which is similar to \ref. So if the thirteenth bibliographic item is introduced with

```
\bibitem{eM57}
```

then

```
\cite{eM57}
```

refers to that item and typesets: [13]. A bibliography is automatically numbered by LaTeX.

Tip   Do not leave spaces in a \cite command; for example, \cite{eM57␣} will produce [?] for an unknown reference.

You can use \cite in the form:

```
\cite{hA70,eM57}
```

(or more than two labels) which typesets as [1, 13]. There is also an optional argument for \cite:

```
\cite[pages~2--15]{eM57}
```

which prints [13, pages 2–15].

If you wish to use labels, specify these with an optional argument of \bibitem:

[EM57] Ernest T. Moynahan, *On a problem of M. H. Stone*, Acta Math. Acad. Sci. Hungar. **8** (1957), 455–460.

typed as

```
\bibitem[EM57]{eM57}
  Ernest~T. Moynahan, \emph{On a problem of M.~H. Stone},
  Acta Math. Acad. Sci. Hungar. \textbf{8} (1957), 455--460.
```

If this optional argument of \bibitem is used, then the \cite command will produce [EM57]. Make sure that the argument of \begin{thebibliography} is wide enough to accommodate all the labels.

Rule ■ Label for a bibliographic item
A label cannot start with a space and cannot contain a comma or a space.

The examples follow the formatting rules of the $\mathcal{AMS}$. Only the titles are emphasized, and only the volume numbers of the journals are set in boldface. Otherwise, just watch the order in which the items are given, the punctuation, and the capitalization.

If an author appears repeatedly as the author of a bibliographic item, use the \bysame command, which replaces the author's name with a long dash _____ followed by a thin space. Example:

```
\bibitem{gF86}
  Grant~H. Foster,
  \emph{Computational complexity in lattice theory},
  tech. report, Carnegie Mellon University, 1986.
\bibitem{gF86a}
  \bysame,
  \emph{Computational complexity in lattice theory},
  Research Note 128A, Carnegie Mellon University,
  Pittsburgh PA, December 1986, research article in preparation.
```

See sampart.tex (page 371) for another example.

The standard LATEX document classes do not provide the \bysame command (the $\mathcal{AMS}$ document classes do), so it's included in the lattice.sty command file (see section 9.5):

```
\providecommand{\bysame}{\makebox[3em]{\hrulefill}\thinspace}
```

See section 9.1.3 for the `\providecommand` command.

Tip   If you want a different title for the bibliography, say, "My title", place the command

```
\renewcommand{\bibname}{My title}
```

anywhere before the `thebibliography` environment (see section 9.1.5).

Tip   You may have more than one `thebibliography` environment in an article. In each bibliography the entries will be numbered from 1.

6.5.2   Index

Using the `\label` and `\pageref` commands (see section 6.4.2), it's quite simple to produce a small index in a `theindex` environment. At each point in the text that you want to reference in the index, place a `\label` command. This will be referenced in the index with a `\pageref`.

Place the index entries in the `theindex` environment. The `\item`, `\subitem`, `\subsubitem` commands introduce an entry, subentry, and subsubentry, respectively. If you require vertical spacing when the first letter changes (say, between the "h" entries and the "i" entries), use the `\indexspace` command.

In the example below the `\com` command is used, which is defined as follows:

```
\newcommand{\com}[1]{\texttt{\symbol{92}#1}}
```

Then `\com{textit}` prints `\textit`.

Here are some examples of index entries. (The labels used in these examples are for illustration only; they have no intrinsic meaning.)

```
\begin{theindex}
  \item \com{AA} (\AA), \pageref{aa1}, \pageref{aa2}
  \item \com{aa} (\aa), \pageref{aa3}, \pageref{aa4}
  \item accent
    \subitem European, \pageref{accentfor1},
        \textbf{\pageref{accentfor2}},
        \pageref{accentfor3}, \textbf{\pageref{accentfor4}}
    \subitem math, \pageref{mthacc1}, \textbf{\pageref{mthacc2}},
        \textbf{\pageref{mthacc3}}
\end{theindex}
```

The typeset index is similar to the index at the end of this book. For a large index, use the *MakeIndex* program (see Chapter 11).

6.6   *Page style*

In this chapter, we have discussed the logical design of a document; the visual design is largely left to the document class. But there is one small aspect of the visual design we have to discuss: the page style.

To get a visual representation of the page style of your document, use the layout package of Kent McPherson. Invoke the package with

```
\usepackage{layout}
```

and place the \layout command in the body of the article. This produces a picture, such as that shown in Figure 6.4 for the article document class with no options. In case of two-sided printing, two pictures are produced.

A typeset page has three parts: the *header* (or running head), the *body* of the page, and the *footer*. As a rule, the document class will take care of the contents and formatting of the header and footer, and the formatting of the body.

You can override the page design of the document class with the command

```
\pagestyle{style}
```

where the argument *style* is one of the following:

- plain, which makes the header empty and the footer containing only the page number;
- empty, which makes both the header and the footer empty;
- headings, which makes the header contain the information provided by the document class and the footer empty;
- myheadings, which makes the header contain the information provided by the

```
\markboth    and    \markright
```

commands and makes the footer empty.

The \markright command has only one argument; the last \markright on a page provides the header information for that page. The \markboth command has two arguments: the first provides the header information for the left-hand page, and the second provides the header information for the right-hand page.

The \thispagestyle command is the same as \pagestyle except that it applies only to the current page.

For instance, if the current page is a fullpage picture, you may want to issue the command

```
\thispagestyle{empty}
```

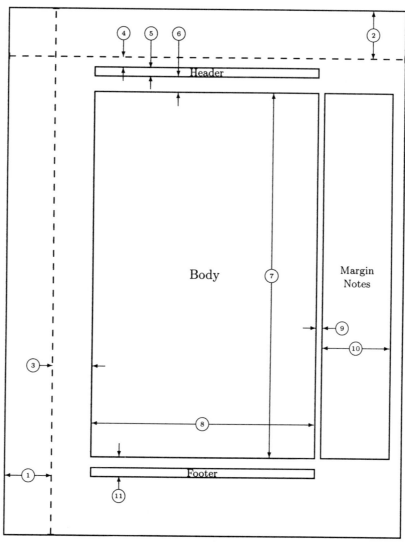

Figure 6.4: Page layout for the article document class

The \maketitle command automatically issues a

\thispagestyle{plain}

command, so if you want to suppress the page number on the first page of an article, you have to put

\thispagestyle{empty}

immediately *after* the \maketitle command.

All of Chapter 4 of *The LATEX Companion* deals with the layout of the page. There is an excellent package, fancyheadings, by Piet van Oostrum for creating your own page style. Appendix G should assist you in getting this package (see also [18]).

7

Standard LATEX document classes

LATEX is a markup language. The marks (commands) are translated by LATEX, by the document class, by the packages, and by the user-defined macros to produce a typeset document.

In this chapter, we discuss the standard LATEX document classes. The choice of the document class strongly affects the appearance of the typeset copy; compare the article typeset with the amsart document class on pages 361–363 with the article typeset with the article document class on pages 39–40.

7.1 The article, report, and book document classes

Most LATEX documents are written with one of the three standard LATEX document classes: article, report, or book (or variants thereof). In sections 1.5 to 1.7, you were shown how to write an article with the article document class. To see how the report and book document classes differ, typeset the article intrart.tex in these two document classes, that is, change the argument of the

```
\documentclass{article}
```

command first to report and then to book. There are two substantive differences you should remember.

- The `report` and `book` document classes have an additional sectioning command: `\chapter` (between `\part` and `\section` in the hierarchy).
- The `book` document class defines no abstract environment.

The markup rules for the three document classes are more or less the same, but the appearance of the same document typeset with the three document classes is substantially different. For instance, by default, the `report` document class provides a separate page for the abstract, but the `article` document class does not.

The book document class formalizes the division of the body into three parts: the front matter, the main matter, and the back matter, introduced with the commands: `\frontmatter`, `\mainmatter`, and `\backmatter`.

The front matter of a book is introduced with the `\frontmatter` command. It normally contains the title page(s), copyright page, preface, foreword, the table of contents, and related matters (as a rule, books do not have abstracts). LᴬTEX numbers the pages with roman numerals. In the front matter the `\chapter` command does not produce a chapter number but the title is listed in the table of contents. So you can start the introduction with

```
\chapter{Introduction}
```

You should use the *-ed form of the other sectioning commands.

The `\mainmatter` command introduces the main matter; page numbering starts from 1 using arabic numerals.

Finally, the back matter is introduced with the `\backmatter` command. It normally contains the afterword, index, colophon, and related matters. The behavior of the `\chapter` command is the same as in the front matter.

The `\maketitle` command is not often used with the book document class.

7.1.1   *More on sectioning*

All sectioning commands take one of the following three forms, illustrated below with the `\section` command:

Form 1   The simplest form is

```
\section{title}
```

where *title* is the section title, of course.

Form 2   The sectioning command may have an optional argument:

```
\section[short_title]{title}
```

The optional *short_title* argument is used in the running head and table of contents.

Form 3 Finally, the ∗-ed version:

`\section*{title}`

There are no sectioning numbers printed and the `title` is used neither in the running head nor in the table of contents.

7.1.2 *Options*

A document's appearance is primarily determined by the document class; however, it can be much influenced by options. The three document classes discussed in this section have eight standard options, affecting many attributes. In this section, we discuss all the options recognized by the standard LaTeX document classes. Combining the three document classes with the various options allow you to print more than 5,000 versions of the same document.

For each attribute there is a *default value*, the one in effect if no option is invoked.

Font size

Options:	10pt
	11pt
	12pt
Default:	10pt

The three options declare the named size the default font size. You may want to use the 12pt option for proofreading:

`\documentclass[12pt]{article}`

However, you must realize that changing the font size changes the line breaks, so changing the 12pt option back to 10pt may require some adjustment.

Paper size

Options:	letterpaper	(8.5 inches by 11 inches)
	legalpaper	(8.5 inches by 14 inches)
	executivepaper	(7.25 inches by 10.5 inches)
	a4paper	(210 mm by 297 mm)
	a5paper	(148 mm by 210 mm)
	b5paper	(176 mm by 250 mm)
Default:	letterpaper	

Draft

Options: draft
            final
Default: final

The draft option prints a slug on the margin next to each line that is too wide. The final option does not.

Two-sided printing

Options: twoside
            oneside
Default: depends on the document class

The twoside option formats the output for printing on both sides of the paper. The default is oneside for the article and the report document classes, and twoside for the book document class.

Chapter start

Options: openright
            openany
Default: depends on the document class

A chapter always starts on a new page. The openright option starts chapters on a right-hand page, while the openany option does not. These options do not apply to the article document class (which does not have chapters). The default for the report document class is openany. The default for the book document class is openright.

Two-column printing

Options: twocolumn
            oneside
Default: onecolumn

The twocolumn option prints the document in two-column format.

Title page

Options: titlepage
            notitlepage
Default: depends on the document class

The titlepage option creates a separate title page and places the abstract on a separate page. The notitlepage option creates no separate pages. For the article document class the default is notitlepage, and for the report and book document classes the default is titlepage.

Equations and equation numbers

> *Options:* leqno
> reqno
> *Default:* leqno

The leqno option places the equation numbers on the left side, while reqno places them on the right side. In this book, I use the leqno option, which is the default in the $\mathcal{A}\mathcal{M}\mathcal{S}$ document classes (see section 8.4).

Option: fleqn

The fleqn option left-aligns displayed formulas; this is typically used in conjunction with the reqno option.

Combinations

Of course, these options can be combined with each other and with most document classes. For instance,

\documentclass[12pt,a4paper,twoside,twocolumn]{report}

will produce a report in a 12 point font size with a paper size that is standard outside the USA, formatted for two-sided, two-column printing.

7.2 *The* letter *document class*

The letter document class was developed for writing letters. One document can contain any number of letters, each in its own letter environment. In the following example (letter.tex in the ftp directory) there is only one letter:

```
% Sample file: letter.tex
% Typeset with LaTeX format
\documentclass{letter}

\begin{document}

\address{George Gr\"{a}tzer\\
        Department of Mathematics\\
        University of Manitoba\\
        Winnipeg, MB, R3T 2N2\\
        Canada}
\signature{George Gr\"{a}tzer}
\date{}
```

```
\begin{letter}{Prof.~John Hurtig\\
                Computer Science Department\\
                University of Winnebago\\
                Winnebago, Minnesota 23714}
\opening{Dear John,}
Enclosed you will find the first draft of the five year plan.
\closing{Friendly greetings,}
\cc{Carla May\\
    Barry Bold}
\encl{Five year plan}
\ps{P.S. Remember our lunch meeting tomorrow. G.}
\end{letter}


\end{document}
```

The argument of the `letter` environment is the name and address of the recipient. It is a required argument, if omitted you get an error message:

```
! Incomplete \iffalse; all text was ignored after line 21.
<inserted text>
                \fi
l.21 \end{letter}
```

In multiline arguments, the lines are separated by \\.

The arguments of some commands may apply to all the letters. Such commands should be placed before the first `letter` environment. In the example, `\signature` and `\address` are so placed.

If the `\date` is absent, today's date will be typeset. If you want no date, use an empty argument `\date{}` as in the example. If you want all letters with the same date, the `\date` command should precede the first `letter` environment. Many of the options listed in section 7.1.2 can also be invoked for the `letter` document class.

7.3   The LaTeX distribution

The LaTeX distribution contains a large number of document classes and packages that you probably received from your TeX software supplier. If not, see Appendix G on how to get them.

There are nine document classes. Of the five not discussed in this book, two may be of interest to the general user: the `slides` document class for preparing slides and the `proc` document class for conference proceedings.

A number of packages are distributed with LaTeX; the following should be especially interesting for readers of this book:

- The latexsym package contains the definitions of some symbols (see Appendix A).
- The alltt package provides the `alltt` environment, which is like the `verbatim` environment except that \, {, and } have their usual meanings. This environment is excellent for writing up computer dialogues like those in Appendix G.
- The exscale package provides scaled versions of the math extension font.
- The makeidx package provides commands for producing indexes (see Chapter 11).
- The showidx package causes the index entries to be printed on the margin (see Chapter 11).
- The syntonly package is used to process a document to check its syntax without typesetting it.

The `nfssfont.tex` file in the LaTeX distribution provides font tables for use with the \symbol command (see section 2.4.4).

There are some major software distributions related to LaTeX, including:

- The $\mathcal{AMS}$-LaTeX package, which is discussed in detail in this book.
- The babel package, which supports typesetting in several languages.
- The graphics distribution (mentioned in section 6.4.3), which provides support for the inclusion and transformation of graphics and for typesetting in color. It depends on your printer and printer driver whether you can use the packages in this distribution.
- The psnfss distribution, which provides support for typesetting with a large range of PostScript fonts (see Appendix F).
- The tools distribution—written by the LaTeX3 project team—which is discussed in the next section.

All these distributions come with their own documentation. They are also described in *The LaTeX Companion*.

7.3.1   Tools

Some of these packages are so important that they could well have been incorporated into LaTeX proper. Here is a brief listing:

- The array package contains extended versions of the `array` and `tabular` environments with many extra features.
- The dcolumn package provides alignment on "decimal points" in tabular entries. It requires the `array` package.
- The delarray package adds "large delimiters" around arrays. Requires `array`.
- The hhline package provides horizontal line control in tables.
- The longtable package helps to create multipage tables. It does not require `array`, but it uses its extended features if both are used together.
- The tabularx package defines a variant of the `tabular` environment. It requires the `array` package.

- The afterpage package implements the \afterpage command that causes the commands specified in its argument to be expanded after the current page is output.
- The enumerate package extends the enumerate environment.
- The ftnright package places all footnotes in the right-hand column in documents printed in two columns.
- The indentfirst package indents the first paragraph of sections.
- The layout package shows the page layout defined by the document class; see section 6.6.
- The multicol package provides multicolumn typesetting.
- The showkeys package selectively prints the labels used by \label, \ref, \cite, and so on (see section 6.4.2).
- The theorem package allows the definition of flexible proclamations; an *AMS*-LaTeX variant, the amsthm package, is discussed in section 3.4.2.
- The varioref package provides "smart" (and multilingual) handling of page references.
- The verbatim package extends the verbatim environment and introduces the indispensable comment environment.
- The xr package allows cross-references between documents.
- The xspace package provides a "smart space" command that helps you avoid the common mistake of missing spaces after command names; to be used mainly in text (see section 9.1.1).

fontsmpl is a test file to produce font samples.

These packages come with documentation; for instance, to get the documentation for the showkeys package, typeset the showkeys.dtx file with LaTeX format. All these packages are discussed in *The LaTeX Companion*.

8

$\mathcal{A}_{\mathcal{M}}\mathcal{S}$-L#ATEX documents

In the Introduction, $\mathcal{A}_{\mathcal{M}}\mathcal{S}$-L#ATEX was defined as a collection of $\mathcal{A}_{\mathcal{M}}\mathcal{S}$ enhancements to L#ATEX. In this chapter, we discuss the $\mathcal{A}_{\mathcal{M}}\mathcal{S}$-L#ATEX document classes and the $\mathcal{A}_{\mathcal{M}}\mathcal{S}$-L#ATEX packages.

The $\mathcal{A}_{\mathcal{M}}\mathcal{S}$-L#ATEX document classes are introduced section 8.1. In section 8.2, the rules of the top matter are reviewed. By following the steps in section 8.3, you'll create an article template for the amsart document class that will hopefully serve your needs. A document class is shaped by its options; in section 8.4, we discuss the options of the $\mathcal{A}_{\mathcal{M}}\mathcal{S}$ document classes. Section 8.5 briefly describes the various packages comprising $\mathcal{A}_{\mathcal{M}}\mathcal{S}$-L#ATEX and their interdependencies.

8.1 The three $\mathcal{A}_{\mathcal{M}}\mathcal{S}$ document classes

The $\mathcal{A}_{\mathcal{M}}\mathcal{S}$-L#ATEX distribution comes with three document classes: amsart ($\mathcal{A}_{\mathcal{M}}\mathcal{S}$ article document class), amsbook ($\mathcal{A}_{\mathcal{M}}\mathcal{S}$ book document class), and amsproc ($\mathcal{A}_{\mathcal{M}}\mathcal{S}$ proceedings document class). The three document classes have many similarities and some differences. The markup for the three document classes is similar, but the actual appearance of a document typeset with the three document classes differs substantially.

The major $\mathcal{A}\mathcal{M}\mathcal{S}$ document class is amsart; most of this chapter deals with it. The amsart document class is popular because it produces typeset articles that look very professional. Compare the sample article with the amsart document class on pages 361–363 to the sample article with the article document class on pages 39–40.

The amsbook document class is not meant to be for books as amsart is for articles; while amsart provides complete formatting for articles, amsbook only lays the foundation for the formatting of a book. amsbook provides the additional sectioning \chapter command but does very little with the front and back matter.

The amsproc document class is for proceedings of meetings; more detailed instructions should come from the editor of the proceedings volume.

The document instr-1.tex in the $\mathcal{A}\mathcal{M}\mathcal{S}$ distribution discusses the $\mathcal{A}\mathcal{M}\mathcal{S}$ document classes. Note a small difference between the structure of $\mathcal{A}\mathcal{M}\mathcal{S}$-L*A*T*E*X and of L*A*T*E*X document classes: the $\mathcal{A}\mathcal{M}\mathcal{S}$-L*A*T*E*X document classes require that you place the abstract *before* the

```
\maketitle
```

command.

8.1.1   Font size commands

In the $\mathcal{A}\mathcal{M}\mathcal{S}$ document classes, in addition to the L*A*T*E*X font size commands (see section 2.6.8), there are five font size commands below \normalsize:

\Tiny

\tiny

\SMALL (a synonym for \scriptsize)

\Small (a synonym for \footnotesize)

\Small

So the complete sequence is:

\Tiny	\tiny	\SMALL	\Small	\small
		\normalsize		
\large	\Large	\LARGE	\huge	\Huge

Two commands allow the user to move up or down this ladder: \larger moves up one and \smaller moves down one. Both take an optional argument; for example, \larger[2] moves up 2.

8.2   The top matter

For a fairly representative example, see the top matter of the sampart.tex sample article on page 361. The title page information is provided as the arguments of several commands. For the your convenience, I'll divide them into three groups.

There is only one general rule:

Rule ■ Top matter commands
All top matter commands are short.

This means that there can be no blank line in the argument of any of these commands (section 2.3.3).

8.2.1 *Article info*

Rule ■ Title

- Command: \title
- Separate lines with \\
- Optional argument: short title for running head
- Do not put a period at the end of a title.

Many titles are too long to be typeset in one line in the large type used by the amsart document class, so indicate where the line should be broken with \\ commands.

The *running head* (the top line of the page, or "header") is the title on odd-numbered pages, set in capital letters. If the title is more than a few words long, use an optional argument to specify a short title for the running head; do not use \\ in the short title.

Examples. A title:

```
\title{A construction of distributive lattices}
```

A title and a short title (for use in the running head):

```
\title[Complete-simple distributive lattices]
    {A construction of complete-simple\\
    distributive lattices}
```

Rule ■ Translator

- Command: \translator
- Do not put a period at the end of the argument.

Example:

```
\translator{Harry~M. Goldstein}
```

Rule ■ Dedication

- Command: \dedicatory
- Separate lines with \\

It is suggested that you indicate to LaTeX where the line should be broken. Example:

```
\dedicatory{To the memory of my esteemed
    friend and teacher,\\ Harry~M. Goldstein}
```

Rule ■ Date

- Command: \date

Examples:

```
\date{January 22, 1991}
```

You can use the \today command to get today's date.

```
\date{\today}
```

To get no date, give the command \date{} or omit the \date command.

8.2.2 *Author info*

Rule ■ Author

- Command: \author
- Optional argument: short form of name for running head

Examples. An author:

```
\author{George~A. Menuhin}
```

An author with a short form of the name for the running head:

```
\author[G.~A. Menuhin]{George~A. Menuhin}
```

Section 8.2.4 deals with multiple authors.

Rule ■ Address

- Command: \address
- Separate lines with \\
- Optional argument: name of author

Example:

> DEPARTMENT OF APPLIED MATHEMATICS, UNIVERSITY OF WINNEBAGO, WINNEBAGO, MINNESOTA 23714

typed as

```
\address{Department of Applied Mathematics\\
         University of Winnebago\\
         Winnebago, Minnesota 23714}
```

Observe that the $\mathcal{AMS}$ document class replaces the \\ line separators with commas.

If there are several authors, it may not be clear how to associate the addresses with the authors. In such cases, use the optional argument of \address (the author's name) to avoid ambiguity. See Example 4 in section 8.2.5 (page 252) for a complete example.

Rule ■ Current address

- Command: \curraddr
- Separate lines with \\
- Optional argument: name of author

Example:

> *Current address*: Department of Mathematics, University of York, Heslington, York, England

typed as

```
\curraddr{Department of Mathematics\\
          University of York\\
```

```
Heslington, York, England}
```

If there are several authors, it may not be clear how to associate the current addresses with the corresponding authors. In such cases, use the optional argument (the author's name) of \curraddr to avoid ambiguity; for some examples, see section 8.2.5.

Rule ■ E-mail

- Command: \email
- Optional argument: name of author

Example:

```
\email{gmen@ccw.uwinnebago.edu}
```

Tip   Some e-mail addresses contain the % character; recall (section 2.4.4) that you have to type \% to get %.

Example:

```
\email{h1175moy\%ella@relay.eu.net}
```

Tip   Some "generic" e-mail addresses contain the special underscore character (_). Recall (section 2.4.4) that you have to type _ to get _.

Example:

```
\email{George\_Gratzer@umanitoba.ca}
```

Rule ■ Research support or other acknowledgment

- Command: \thanks
- Do not indicate linebreak.

Example:

```
\thanks{Research was supported in part by NSF grant PAL-90-2466.}
```

8.2.3 $\mathcal{A_MS}$ info

The following are collected at the bottom of the first page as unmarked footnotes.

Rule ■ $\mathcal{A_MS}$ subject classification

- Command: \subjclass
- The amsart document class adds a period at the end.
- The argument should be either a five-digit code, or the string Primary: followed by a five-digit code, a semicolon, the string Secondary:, and another five-digit code.

Examples:

```
\subjclass{06B10}
\subjclass{Primary: 06B10; Secondary: 06D05}
```

The current subject classification scheme is available electronically from the $\mathcal{A_MS}$ (see Appendix G on how to access it).

Rule ■ Keywords

- Command: \keywords
- Do not indicate line breaks.
- The $\mathcal{A_MS}$ document class supplies the phrase "*Key words and phrases.*" and a period at the end of the subject classification.

Example:

```
\keywords{Complete lattice, distributive lattice, complete
    congruence, congruence lattice}
```

Further footnotes An additional \thanks becomes an additional footnote. Examples:

```
\thanks{This is a preliminary version of this article,
        prepared for the Second Annual Meeting of the
        Statistical Association of Winnebago.}
\thanks{This article is in final form, and no version of it
        will be submitted elsewhere.}
```

The second example may be used in conference proceedings to indicate that the article should be reviewed.

8.2.4   *Multiple authors*

If the article has several authors, the author information should be repeated for each one. Take care that the e-mail follows the address.

In case two authors have the same address, omit the \address for the second author (who can still have \email). In this case, an additional footnote should be a \thanks following the \thanks of the *last author*. Since the footnotes are not marked, the argument of the \thanks for research support should contain a reference to the author:

```
\thanks{The research of the first author was supported in part by
        NSF grant PAL-90-2466.}
\thanks{The research of the second author was supported by
        the Hungarian National Foundation for Scientific Research,
        under Grant No.~9901.}
```

Finally, if an article has so many authors that even the short form of the authors' name is too long for the running head, use only one \author command as in the following example:

```
\author[G.~A. Menuhin, E.~T. Moynahan, et al.]
    {George~A. Menuhin, Ernest~T. Moynahan, Robert~S. Treblinski,
    Pavel~G. Viznobranski, and Berry~R. Wojdicko}
```

Now write the author info as usual, but without any more \author commands. So for the second, third, and so on authors, only type \address, \email, and \thanks.

If there are multiple authors, sometimes it may not be clear whose address, current address, or e-mail address is being given. In such cases give the name of the author as an optional argument. Example:

> *Email address,* Ernest T. Moynahan: emoy@ccw.uwinnebago.edu.

typed as

```
\email[Ernest~T. Moynahan]{emoy@ccw.uwinnebago.edu}
```

See also Example 4 in section 8.2.5.

8.2.5   *Examples*

Here are some examples of the top matter; they are contained in the topmat.tpl file in the ftp directory—see page 4.

Example 1 One author:

```
% Article info
\title[Complete-simple distributive lattices]
        {A construction of complete-simple\\
         distributive lattices}
\date{\today}

% Author info
\author{George~A. Menuhin}
\address{Computer Science Department\\
         University of Winnebago\\
         Winnebago, Minnesota 23714}
\email{gmen@ccw.uwinnebago.edu}
\thanks{This research was supported by
        the NSF under grant number~23466.}

% AMS info
\keywords{Complete lattice, distributive lattice,
          complete congruence, congruence lattice}
\subjclass{Primary: 06B10; Secondary: 06D05}
```

In \title, the optional argument (the running head) is the rule, not the exception. All the items are required, except \email.

Example 2 Two authors. I only show the author info section; the others are unchanged from Example 1:

```
% Author info
\author{George~A. Menuhin}
\address{Computer Science Department\\
         University of Winnebago\\
         Winnebago, Minnesota 23714}
\email{gmen@ccw.uwinnebago.edu}
\thanks{The research of the first author was
        supported by the NSF under grant number~23466.}
\author{Ernest~T. Moynahan}
\address{Mathematical Research Institute
         of the Hungarian Academy of Sciences\\
         Budapest, P.O.B. 127, H-1364\\
         Hungary}
\email{h1175moy\%ella@relay.eu.net}
\thanks{The research of the second author
        was supported by the Hungarian
        National Foundation for Scientific Research,
        under Grant No.~9901.}
```

Example 3   Two authors, same department. I only show the author info section; the others are unchanged from Example 1:

```
%  Author info
\author{George~A. Menuhin}
\address{Computer Science Department\\
        University of Winnebago\\
        Winnebago, Minnesota 23714}
\email[George~A. Menuhin]{gmen@ccw.uwinnebago.edu}
\thanks{The research of the first author was
        supported by the NSF under grant number~23466.}
\author{Ernest~T. Moynahan}
\email[Ernest~T. Moynahan]{emoy@ccw.uwinnebago.edu}
\thanks{The research of the second author
        was supported by the Hungarian National
        Foundation for Scientific Research,
        under Grant No.~9901.}
```

Example 4   Three authors, the first two from the same department, the second and third with e-mail addresses and research support. I only show the author information section; the others are unchanged. There are various ways of doing this, here is one possibility:

```
%  Author info
\author{George~A. Menuhin}
\address[George~A. Menuhin and Ernest~T. Moynahan]
   {Computer Science Department\\
    University of Winnebago\\
    Winnebago, Minnesota 23714}
\author{Ernest~T. Moynahan}
\email[Ernest~T. Moynahan]{emoy@ccw.uwinnebago.edu}
\thanks{The research of the second author was
        supported by the Hungarian National
        Foundation for Scientific Research,
        under Grant No.~9901.}
\author{Ferenc~R. Richardson}
\address[Ferenc~R. Richardson]
   {Department of Mathematics\\
    California United Colleges\\
    Frasco, CA 23714}
\email{frich@ccu.frasco.edu}
\thanks{The research of the third author was
        supported by the NSF under grant number~23466.}
```

Tip   The most typical mistake in the top matter is the misspelling of a command name, for instance, \adress. This is no problem because the error message:

```
! Undefined control sequence.
1.37 \adress
                {Computer Science Department\\
```

is very helpful. Similarly, if you drop a closing brace, for instance:

```
\email{menuhin@ccw.uwinnebago.edu
```

you are told what went wrong, since all these commands are short, see section 2.3.3:

```
Runaway argument?
{menuhin@ccw.uwinnebago.edu \thanks {The research of th\ETC.
! Paragraph ended before \email was complete.
<to be read again>
                        \par
1.52
```

If you drop an opening brace:

```
\author George~A. Menuhin}
```

you get the error message:

```
! Too many }'s.
1.43 \author George~A. Menuhin}
```

If you enclose an optional argument by braces, say,

```
\title{Complete-simple distributive lattices}%
      {A construction of complete-simple\\
        distributive lattices}
```

the amsart document class makes the short title into the title (the real title is printed before it).

8.3   A𝓜S article template

In this section, you'll create a template for your A𝓜S articles, using the amsart document class. It'll contain a customized preamble, preformatted top matter, and sample bibliographic items. A template is a file that is on disk as a "read-only" file. With an editor, open the file, save it under a new name (say, in the work subdirectory), and proceed to write the new article—without having to remember the details governing the preamble and the top matter.

You'll create the template in several steps.

Step 1   In an editor, open the document `amsart.tpl` in the `ftp` directory (see page 4, or type in the lines as shown in this section) and save it in the `work` subdirectory under the name `myams.tpl`.

The first few lines read:

```
% Sample file: amsart.tpl
% Typeset with LaTeX format

% Preamble
\documentclass{amsart}
\usepackage{amssymb}
```

Use commented out lines (lines that start with %) for comments.

Rewrite line 1 to read:

```
% Sample file: myams.tpl
```

The lines

```
\documentclass{amsart}
\usepackage{amssymb}
```

specify the amsart document class, and the use of the amssymb package to gain access to all the symbols listed in Appendix A by name.

Step 2   After the \usepackage command, five options are presented that correspond to the five example sets of proclamation definitions in section 3.4.2.

Choose Option 5 for `myams.tpl` by deleting all the lines related to the other options. You are left with the lines:

```
% theorems, corollaries, lemmas, and propositions,
% in the most emphatic (plain) style;
% all are numbered separately
% There is a Main Theorem in the most emphatic (plain)
% style, unnumbered
% There are definitions, in the less emphatic (definition) style
% There are notations, in the least emphatic (remark) style,
% unnumbered

\theoremstyle{plain}
\newtheorem{theorem}{Theorem}
\newtheorem{corollary}{Corollary}
\newtheorem*{main}{Main Theorem}
\newtheorem{lemma}{Lemma}
\newtheorem{proposition}{Proposition}
```

```
\theoremstyle{definition}
\newtheorem{definition}{Definition}

\theoremstyle{remark}
\newtheorem*{notation}{Notation}
```

Step 3 Then two choices are presented: one author or two authors (for more complicated situations, see section 8.2.4). For the template `myams.tpl`, choose one author by deleting everything between the lines (inclusive)

```
%   Two authors
```

and

```
%   End Two authors
```

You are left with:

```
\begin{document}
% One author
\title[shorttitle]{titleline1\\
                    titleline2}
\author{name}
\address{line1\\
        line2\\
        line3}
\email{name@address}
\thanks{thanks}
% End one author

\keywords{keywords}
\subjclass{Primary: subject; Secondary: subject}
\date{date}

\begin{abstract}
   abstract
\end{abstract}
\maketitle

\begin{thebibliography}{99}

\end{thebibliography}
\end{document}
```

In the top matter, provide your own personal information. For instance, I edit

```
\author{name}
```

to read

```
\author{George~Gr\"{a}tzer}
```

and I also edit \address, \email, and \thanks.
     After editing, I get:

```
% top matter
\title[shorttitle]{titleline1\\
                   titleline2}
\author{George~Gr\"{a}tzer}
\address{University of Manitoba\\
        Department of Mathematics\\
        Winnipeg, MN, R3T 2N2\\
        Canada}
\email{George\_Gratzer@umanitoba.ca}
\thanks{Research supported by the NSERC of Canada.}

\keywords{keywords}
\subjclass{Primary: subject; Secondary: subject}
\date{date}

\begin{abstract}
   abstract
\end{abstract}
\maketitle

\begin{thebibliography}{99}

\end{thebibliography}
\end{document}
```

This is a template for all future articles, so I do not edit the lines (\title, \keyword, and so on) that change from article to article; I leave them generic.
     Note that the short title is for "running heads" (also called "headers"), that is, for the title shown at the top of every odd-numbered page other than the title page; if the title is only one line long, delete the separation mark \\ and the second line (except for the closing brace). If the title is short, delete [shorttitle].
     Now save myams.tpl. You could also make a second version with two authors to be used as a template for joint articles. Note that at the end of the template, just before the line \end{document}, there are two lines:

```
\begin{thebibliography}{99}
```

`\end{thebibliography}`

The argument of `\begin{thebibliography}` should be 9 if there are fewer than 10 references; it should be 99 with 10–99 references, and so forth. In sections 1.7.4 and 6.5.1, we discussed how to insert bibliographic items; the amsart document class does not change these rules. The templates for the bibliographic items are listed after the line `\end{document}` as shown in section 1.7.4.

To make sure that you do not overwrite your template, make it "read-only"; how you do this, depends on the computer system you are using.

8.4 *Options*

The $\mathcal{A_MS}$ document classes support a number of options, affecting many attributes. For each attribute there is a *default value*, the one in effect if no option is invoked.

Font size

Options:	8pt
	9pt
	10pt
	11pt
	12pt
Default:	10pt

These options declare the default font size. You may want to use the 12pt option for proofreading:

`\documentclass[12pt]{amsart}`

However, you must realize that changing the font size changes the line breaks, so changing the 12pt option back to 10pt may require some adjustment.

Paper size

Options:	letterpaper	(8.5 inches by 11 inches)
	legalpaper	(8.5 inches by 14 inches)
	a4paper	(210 mm by 297 mm)
Default:	letterpaper	

Draft

Options:	draft
	final
Default:	final

$$\int_0^\pi \sin x \, dx = 2 \tag{1}$$

Figure 8.1: `fleqn` and `reqno` options for equations

$$
\begin{aligned}
(1) \qquad f &= (x_1 x_2 x_3 x_4 x_5 x_6)^2 \\
&= (x_1 x_2 x_3 x_4 x_5 + x_1 x_3 x_4 x_5 x_6 + x_1 x_2 x_4 x_5 x_6 + x_1 x_2 x_3 x_5 x_6)^2 \\
&= (x_1 x_2 x_3 x_4 + x_1 x_2 x_3 x_5 + x_1 x_2 x_4 x_5 + x_1 x_3 x_4 x_5)^2
\end{aligned}
$$

Figure 8.2: Top-or-bottom tags option for `split`

The `draft` option prints a slug on the margin next to each line that is too long. The `final` option does not.

Equations and equation numbers

A number of options deal with the placement of equations and equation numbers.

> *Options:*   `leqno`
>   `reqno`
> *Default:*   `leqno`

By default, equation numbers are placed on the left (the `leqno` option). The `reqno` option places the equation numbers on the right. Note that this is the reverse of the LATEX default (section 7.1.2).

Option: `fleqn`

This option positions equations a fixed amount of space from the left margin rather than centering them; the `fleqn` option is typically used used in conjunction with the `reqno` option. The `fleqn` and `reqno` options are jointly illustrated in Figure 8.1.

> *Options:*   `tbtags`
>   `centertags`
> *Default:*   `centertags`

The `tbtags` option uses "top-or-bottom tags" for a `split` environment; that is, it places the equation number level with the last line, if numbers are on the right, and with the first line, if the numbers are on the left; this option is illustrated in Figure 8.2. The `centertags` option (the default) vertically centers the equation number for a `split` environment.

Two-sided printing

Options: twoside
          oneside
Default: twoside

The twoside option formats the output for printing on both sides of the paper; the opposite is the oneside option. This option influences headers, page numbers, and so on.

Two column printing

Options: twocolumn
          oneside
Default: onecolumn

The twocolumn option prints the document in two columns.

Titlepage

Options: titlepage
          notitlepage
Default: depends on the document class

The titlepage option creates a separate title page and also places the abstract on a separate page. The notitlepage option creates no separate pages.

The default depends on the document class; for instance, for the amsart document class the default is notitlepage, and for the amsbook document class the default is titlepage.

Fonts

Option: noamsfonts

With this option, the document class does not load the packages necessary for the use of the AMSFonts font set.

Option: psamsfonts

LaTeX has to be told if the Y&Y/Blue Sky PostScript version of the AMSFonts font set is to be used. This is done with the psamsfonts option.

No math

Option: nomath

By default, the $\mathcal{AMS}$ document classes load the amsmath package. If you want to use the title page and related features without the math, then use the nomath option.

8.4.1 *Math options*

There are many options for the amsmath package.

The `fleqn`, `reqno`, `leqno`, `tbtags`, and `centertags` document class options (see section 8.4) are passed on to the amsmath package. They can also be given directly to the package, for instance:

```
\documentclass{article}
\usepackage[leqno]{amsmath}
```

Limits

Options: `intlimits`
               `nointlimits`

Default: `nointlimits`

The `intlimits` option places the subscripts and superscripts of integral symbols in displayed equations above and below rather than to the side; `nointlimits` positions them to the side.

Options: `sumlimits`
               `nosumlimits`

Default: `nosumlimits`

The `sumlimits` option places the subscripts and superscripts of large operators ($\prod$, $\coprod$, $\otimes$, $\oplus$, and so on) above and below. `nosumlimits` positions them to the side (see Table 4.6).

Options: `namelimits`
               `nonamelimits`

Default: `namelimits`

The `namelimits` option places the subscripts and superscripts of operators such as det, inf, lim, max, min, and so on, above and below. `nonamelimits` positions them to the side. (See Tables 4.3 and 4.4.)

PostScript AMSFonts

Option: `cmex10`

The amsmath package has to be told if the Y&Y/Blue Sky PostScript AMSFonts are used; normally this is done with the `psamsfonts` option of the document class that is passed on to amsmath. If the amsmath package is invoked directly, you must use the `cmex10` option; for instance,

```
\documentclass{article}
\usepackage[cmex10]{amsmath}
```

8.5 The AMS-LaTeX packages

The $\mathcal{AMS}$ distribution contains the following packages that can be invoked together or independently:

Math enhancements

- The amsmath package is the primary math enhancement package; it automatically loads the amsgen, amsbsy, amsopn, and amstext packages.
- The amsbsy package provides two commands for the use of bold math symbols: `\boldsymbol` and `\pmb` (see section 4.14.3); it automatically loads the amsgen package.
- The amscd package does simple commutative diagrams (see section 5.7); it automatically loads the amsgen package.
- The amsgen package is an auxiliary package that is never invoked directly.
- The amsintx package is an experimental package for extended integrals and sums (mentioned in section 4.4.5); it automatically loads the amsgen package.
- The amsopn package provides operator names and the `\DeclareMathOperator` command to define new ones (see section 4.7); it automatically loads the amsgen package.
- The amstext package defines the `\text` command and redefines the commands `\textrm`, `\textbf`, and so on, to behave like `\text` (see section 4.5); it automatically loads the amsgen package.
- The amsthm package provides more flexible formatting of proclamations (theorem-like structures) and defines the `proof` environment (see sections 3.4.2 and 3.5); it automatically loads the amsgen package.
- The amsxtra package provides the "sp" math accents (see section 4.9); it automatically loads the amsgen, amsbsy, amsmath, amsopn, and amstext packages.
- The upref package insures that the `\ref` command always produces roman numbers.

AMSFonts

- The amsfonts package contains the basic commands to utilize the AMSFonts; it also defines the `\mathfrak` command for the Euler Fraktur math alphabet (see section 4.14.2). If you use the Y&Y/Blue Sky PostScript AMSFonts font set, invoke this package with the option

  `\usepackage[psamsfonts]{amsfonts}`

- The amssymb package defines the symbol table; it automatically loads amsfonts.
- The eucal package replaces the calligraphic math alphabet with the Euler Script math alphabet (see section 4.14.2); if you invoke it with the option

  `\usepackage[mathscr]{eucal}`

then both the \mathscr and the \mathcal commands are available.

- The eufrak package defines the Euler Fraktur math alphabet (see section 4.14.2).

The three A$\mathcal{M}$S-L*A*T*E*X document classes automatically load the amsmath, amsbsy, amstext, amsopn, amsthm, and amsgen packages from the first group, and the amsfonts package from the second group. So for normal work, it's sufficient to invoke an A$\mathcal{M}$S document class with

```
\documentclass{amsart}% or amsbook, or amsproc
\usepackage{amssymb}
```

The other packages are loaded as necessary.

A typical article with the `article` document class utilizing the A$\mathcal{M}$S enhancements would normally have:

```
\documentclass{article}
\usepackage{amsmath}% math enhancements
\usepackage{amssymb}% AMSFonts and symbols
```

and maybe one or both of the following:

```
\usepackage{amsthm}% proclamations with style
\usepackage{eucal}% Euler Script
```

Note that it's not critical to remember which packages load others automatically. There is no harm done, say, by typing

```
\usepackage{amsmath}
\usepackage{amsbsy}
```

The amsbsy package will be automatically loaded by the amsmath package, and the

```
\usepackage{amsbsy}
```

line will be ignored by L*A*T*E*X.

The interdependence among the various A$\mathcal{M}$S-L*A*T*E*X packages and document classes is illustrated in Figure 8.3. The document classes are represented by rounded rectangles, the AMSFonts packages by dotted rectangles, and the math enhancements by solid rectangles.

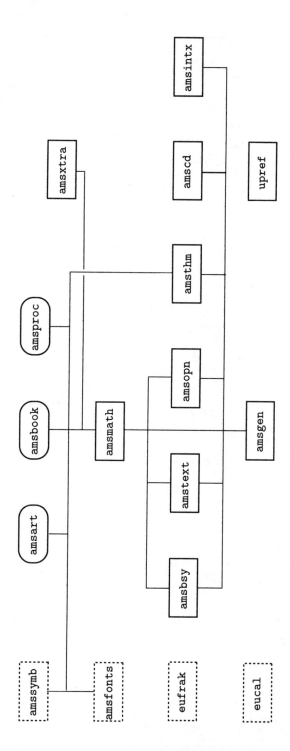

Figure 8.3: *AMS-LᴬTEX package and document class interdependency*

PART IV

Customizing

9

Customizing LaTeX

There is a lot you can do to speed up typing and typesetting in LaTeX. In this chapter, I cover some of the basic techniques.

The chapter starts with user-defined commands and environments (sections 9.1 and 9.2); they are crucial in changing LaTeX to satisfy your particular needs. LaTeX keeps integers it needs to manipulate (equation numbers, section numbers, and so forth) in *counters*; it stores distance measurements (such as \textwidth and \parindent) in *length commands*. In section 9.3, you'll take a closer look at these. Delimited commands, in section 9.4, provide a way of writing the source document in a more readable fashion.

A collection of user-defined commands is presented in section 9.5 as a model to help develop a command file of your own.

Although LaTeX provides three different types of list environments (see section 3.1), it's often necessary to create your own. The list environment, discussed in section 9.6, will do the job.

Finally, in section 9.7, you learn how to make custom formats that speed up the typesetting of a LaTeX document.

9.1   *User-defined commands*

LATEX includes a large number of commands. However, LATEX is much easier to use if you judiciously add *user-defined commands* to satisfy your particular needs.

9.1.1   *Commands as shorthand*

If you use the \leftarrow command a lot, you may want to define:

\newcommand{\la}{\leftarrow}

and then you only have to type \la to obtain a left arrow. Instead of

\widetilde{a}

you can simply type \wa after defining:

\newcommand{\wa}{\widetilde{a}}

I'll show you how to generalize the above command in section 9.1.2.

If you use the construct $D^{[2]} \times D^{[3]}$ often, you may want to introduce

\newcommand{\DD}{D^{[2]}\times D^{[3]}}

and then type \DD instead of D^{[2]} \times D^{[3]} throughout the document.

You can also use commands as a shorthand for text. For instance, if you use the phrase "subdirectly irreducible" many times in your document, you may choose to define

\newcommand{\si}{subdirectly irreducible}

and then "\si" becomes shorthand for "subdirectly irreducible".

Rule   ■   User-defined commands

1. Issue the \newcommand command.
2. In braces, type the new command, including the backslash (\).
3. Again in braces, define the new command.
4. Use new commands like \si above as \si\␣ or \si{} before a space and \si otherwise.

Place user-defined commands in the preamble or in a command (style) file (see section 9.5 for an example of such a file). Then you know where to look for the definition of a command.

Tip   If errors occur, reintroduce new user-defined commands one at a time, so you can isolate the one that causes the problem. LaTeX checks only whether the braces match when the command is read. Other mistakes will not be found until the command is used.

Be careful not to define a user-defined command with a name that is already in use. However, if you do, not much harm is done. You get an error message:

```
! LaTeX Error: Command \la already defined.
```

In this case, you need to *redefine* the command. See section 9.1.3 for how to do this.

Tip   Use spaces to make the source file more readable.

For example, type

```
$D^{ \langle 2 \rangle } + 2 = x^{ \mathbf{a} }$
```

to see more clearly how the braces match, to easily identify relations and operations, and so on. *Do not do this in command definitions, however!* This wastes memory, and may result in unwanted spaces in the typeset document.

Tip   In the definition of a new command, *command declarations* need an extra pair of braces (see section 2.3.3).

Suppose you want to define a command for the warning: *Do not redefine this variable!* It is very easy to make the following error:

```
\newcommand{\Warn}{\itshape Do not redefine this variable!}
```

Then \Warn will present the warning, but everything thereafter will also be italicized. Indeed, \Warn is replaced by

```
\itshape Do not redefine this variable!
```

so the affect of \itshape goes beyond the sentence.

The correct definition is:

```
\newcommand{\Warn}{{\itshape Do not redefine this variable!}}
```

or use a command with an argument:

```
\newcommand{\Warn}{\textit{Do not redefine this variable!}}
```

Tip   The commands that change font characteristics are good candidates for user-defined commands as shorthand (see section 9.5).

However, try not to yield to the temptation to rename them to the old two-letter commands (see section 2.6.6), for instance, \textbf to \bf. There are several reasons for this:

- The old \bf was a command without argument; \textbf is a command with argument.
- Others reading the source document may misunderstand it.
- Some older packages or document classes may use the two-letter commands internally.

Ensuring math

The \ensuremath command is useful for defining commands that work both in text and math mode. Suppose you want to define a shorthand for $D^{(2)}$. If you define it as

```
\newcommand{\Ds}{D^{\langle2\rangle}}
```

then this works in math mode, but not in text. If you define it as

```
\newcommand{\Ds}{$D^{\langle2\rangle}$}
```

then it works in text, but not in math mode. Instead, define this command as follows:

```
\newcommand{\Ds}{\ensuremath{D^{\langle2\rangle}}}
```

so that \Ds will work correctly in all contexts.

This example also shows the editorial advantages of user-defined commands. Suppose a referee suggests that you change the notation to $D^{[2]}$. To carry out the change you only have to change one line:

```
\newcommand{\Ds}{\ensuremath{D^{[2]}}}
```

The xspace *package*

Clause 4 of the rule is the source of many mistakes in LaTeX. David Carlisle's xspace package (see section 7.3.1) helps eliminate such problems. In the preamble, invoke the package with

```
\usepackage{xspace}
```

Whenever you define a command that may have such problems, add to the definition the \xspace command; for instance, define \si as follows

```
\newcommand{\si}{subdirectly irreducible\xspace}
```

Then all of the following will typeset "subdirectly irreducible lattice" correctly:

```
\si\␣lattice
\si{}␣lattice
\si␣lattice
\si␣␣lattice
```

Tip Be careful not to use \xspace in a definition twice.

For instance, if you define

```
\newcommand{\tex}{\TeX\xspace}
\newcommand{\bibtex}{\textsc{Bib}\kern-.1em\tex\xspace}% Bad!!!
```

then

```
\bibtex, followed by a comma
```

prints

BIBTEX , followed by a comma

The correct definitions are

```
\newcommand{\tex}{\TeX\xspace}
\newcommand{\bibtex}{\textsc{Bib}\kern-.1em\TeX\xspace}% Correct!
```

More examples of user-defined commands can be found in section 9.5.

9.1.2 *Arguments*

If you define (using the amsmath package)

```
\newcommand{\Ahh}{\Hat{\Hat{A}}}
```

then you can use the \Ahh command in place of \Hat{\Hat{A}}. If you want to use \Ahh in math and also by itself in text, define it with \ensuremath as

```
\newcommand{\Ahh}{\ensuremath{\Hat{\Hat{A}}}}
```

However, if you use double hats on numerous characters, then you may need a command for double hats in general. Here is how you do it:

`\newcommand{\hh}[1]{\ensuremath{\Hat{\Hat{#1}}}}`

and then to print $\hat{\hat{A}}$, type `\hh{A}`. The form of this `\newcommand` is the same as before, except that after the name of the command `{\hh}` we put in brackets **the number of arguments**, which in this example is [1]. This allows us to use #1 in the definition of the command. When the command is invoked, the argument you provide replaces #1 in the definition. As another example, type `\hh{B}` to get $\hat{\hat{B}}$.

In this section, we present a few simple examples of user-defined commands with arguments.

Example 1   In the preamble of the manuscript of this book, you find

`\newcommand{\env}[1]{\texttt{#1}}`

The `\env` command is used to typeset environment names. So the environment name center is typed as

`\env{center}`

Again the editorial advantage is obvious; if the editor wants the environment names emphasized, just one line in the book has to be changed:

`\newcommand{\env}[1]{\emph{#1}}`

Example 2   The argument #1 may occur more than once in a definition. A natural example is provided by the `\index` command (see section 11.1). Typically, if you want to include a phrase, say, "subdirectly irreducible lattice" in the index, you would have to type

`so this proves that $L$ is a subdirectly irreducible`
`lattice\index{subdirectly irreducible lattice}`

You may wish to introduce the user-defined command

`\newcommand{\ie}[1]{#1\index{#1}}`

where "ie" stands for "index entry". The argument of this command is a phrase in the source file to be included in the index. Using this command, you can simply write

`so this proves that $L$ is a \ie{subdirectly irreducible lattice}`

If you want all index entries in italics, the definition of `\ie` becomes

`\newcommand{\ie}[1]{#1\index{#1@\textit{#1}}}`

in which #1 occurs three times.

Example 3   A user-defined command may have up to nine arguments. Let's define a command with three arguments for congruences:

```
\newcommand{\con}[3]{#1\equiv#2\pod{#3}}
```

Then to print the congruence $a \equiv b \ (\theta)$, type `$\con{a}{b}{\theta}$`. In section 9.4, we present another command for congruences.

Example 4   I mentioned in section 6.4.2 that in this book all sections have labels starting with "`S:`" (see also section 9.5). To refer to a section with label xxx, I type: `section~\ref{S:xxx}`. So I define

```
\newcommand{\refS}[1]{section~\ref{S:#1}}
```

and then the reference is `\refS{xxx}`.

Example 5   This example is from the `sampart2.tex` sample article (see Appendix D). In that article, there are a lot of vectors with one nonzero entry:

$$\langle \dots, 0, \dots, \overset{i}{d}, \dots, 0, \dots \rangle$$

the i on top of d indicates it's the ith component of the vector. In the presence of the amsmath package, a command producing this symbol can be defined by

```
\newcommand{\vct}[2]{\langle\dots,0,\dots,%
\overset{#1}{#2},\dots,0,\dots\rangle}
```

So to print $\langle \dots, 0, \dots, \overset{i}{d}, \dots, 0, \dots \rangle$, type `\vct{i}{d}` in a math formula.

Example 6   In the formula gallery (section 1.3), Formula 20 is the following:

$$\mathbf{A} = \begin{pmatrix} \dfrac{\varphi \cdot X_{n,1}}{\varphi_1 \times \varepsilon_1} & (x+\varepsilon_2)^2 & \cdots & (x+\varepsilon_{n-1})^{n-1} & (x+\varepsilon_n)^n \\ \dfrac{\varphi \cdot X_{n,1}}{\varphi_2 \times \varepsilon_1} & \dfrac{\varphi \cdot X_{n,2}}{\varphi_2 \times \varepsilon_2} & \cdots & (x+\varepsilon_{n-1})^{n-1} & (x+\varepsilon_n)^n \\ \hdotsfor{5} \\ \dfrac{\varphi \cdot X_{n,1}}{\varphi_n \times \varepsilon_1} & \dfrac{\varphi \cdot X_{n,2}}{\varphi_n \times \varepsilon_2} & \cdots & \dfrac{\varphi \cdot X_{n,n-1}}{\varphi_n \times \varepsilon_{n-1}} & \dfrac{\varphi \cdot X_{n,n}}{\varphi_n \times \varepsilon_n} \end{pmatrix} + \mathbf{I}_n$$

This is also a good candidate for user-defined commands:

```
\newcommand{\quot}[2]{%
\dfrac{\varphi \cdot X_{n, #1}}%
{\varphi_{#2}\times \varepsilon_{#1}}}
\newcommand{\exn}[1]{(x+\varepsilon_{#1})^{#1}}
```

So

```
\[
  \quot{2}{3} \qquad \exn{n}
\]
```

prints

$$\frac{\varphi \cdot X_{n,2}}{\varphi_3 \times \varepsilon_2} \qquad (x + \varepsilon_n)^n$$

With these user-defined commands, rewrite Formula 20 as follows:

```
\[
   \mathbf{A} =
   \begin{pmatrix}
      \quot{1}{1} & \exn{2} & \cdots & \exn{n - 1} & \exn{n}\\
      \quot{1}{2} & \quot{2}{2} & \cdots & \exn{n - 1} &\exn{n}\\
      \hdotsfor{5}\\
      \quot{1}{n} & \quot{2}{n} & \cdots &
        \quot{n - 1}{n} & \quot{n}{n}
   \end{pmatrix}
    + \mathbf{I}_{n}
\]
```

Observe how much shorter this form is than the one typed on page 29 and how much easier it's to read.

9.1.3   Redefining commands

LaTeX makes sure that you do not inadvertently define a new command with the same name as an existing command. To test this, define

```
\newcommand{\or}{\vee}
```

Then you get the error message

```
! LaTeX Error: Command \or already defined.
```

```
1.12 \newcommand{\or}{\vee}
```

Assuming that you have already defined the \vct command as in the last section, to *redefine* \vct, use \renewcommand:

```
\renewcommand{\vct}{\langle#1\rangle}
```

Tip   Use \renewcommand sparingly. Make sure that you understand the consequences of redefining an existing command. Redefining LaTeX commands may cause LaTeX to behave in unexpected ways, or even crash.

You can also use \renewcommand to redefine the way LaTeX or any package was programmed to do things. For instance, the end of proof symbol is called \qedsymbol in the amsthm package (automatically loaded by the $\mathcal{AMS}$ document classes). To change this symbol to a symbol some people prefer (defined in the amssymb package), issue the command:

\renewcommand{\qedsymbol}{$\blacksquare$}

Even better, define

\renewcommand{\qedsymbol}{\ensuremath{\blacksquare}}

so that you can use \qedsymbol in both text and math. More on this topic will be found in section 9.1.5.

The \renewcommand command has a companion:

\providecommand

If the command it intends to define is already defined, then \providecommand is ignored; the original definition stays in place. Otherwise, \providecommand acts as a \newcommand. For instance, the \bysame command (see section 6.5.1, page 230), is defined in some document classes as

\newcommand{\bysame}{\makebox[3em]{\hrulefill}\thinspace}

If you want to use the \bysame command in your bibliography, and try to include this definition, LaTeX will indicate an error when you typeset your document using a document class that already contains this definition. Instead, you should define

\providecommand{\bysame}{\makebox[3em]{\hrulefill}\thinspace}

and then the document will typeset correctly whether or not the document class defines the \bysame command.

9.1.4   *Optional arguments*

You can define a command whose first argument is *optional*, and provide a *default value* for this optional argument. To illustrate this, let's define the command:

\newcommand{\Sum}{a_{1}+a_{2}+\cdots+a_{n}}

Then $\Sum$ prints $a_1 + a_2 + \cdots + a_n$. Now let's change this so we can sum from 1 to m if necessary, with n as the default:

\newcommand{\NewSum}[1][n]{a_{1}+a_{2}+\cdots+a_{#1}}

Then $\NewSum$ still prints $a_1 + a_2 + \cdots + a_n$, but $\NewSum[m]$ prints $a_1 + a_2 + \cdots + a_m$.

A \newcommand may have up to nine arguments, but *only the first* may be optional. The following command has two arguments, one optional:

```
\newcommand{\NNsum}[2][n]{#2_{1}+#2_{2}+\cdots+#2_{#1}}
```

and then

`$\NNsum{x}$`	prints	$x_1 + x_2 + \cdots + x_n$
`$\NNsum{a}$`	prints	$a_1 + a_2 + \cdots + a_n$
`$\NNsum[i]{a}$`	prints	$a_1 + a_2 + \cdots + a_i$

9.1.5   Redefining names

A number of phrases, such as "Table", "List of Tables", "Abstract", and so on, are inserted by LaTeX automatically into your typeset document. You can easily change these phrases.

For instance, if for the proceedings of a meeting, a manuscripts is to be submitted in LaTeX, except that "Abstract" has to be changed to "Summary", then do the following. Since "Abstract" is defined with

```
\newcommand{\abstractname}{Abstract}
```

internally, redefine it with

```
\renewcommand{\abstractname}{Summary}
```

Table 9.1 lists the commands that define such names in various LaTeX and AMS-LaTeX document classes, along with their customary meanings.

It is easy to check whether your document class uses a command: just open the cls file of the document class and search for the command. The article, amsart, and amsproc document classes use \refname, while the report, book, and amsbook document classes use \bibname. The amsart and amsbook document classes use a number of additional commands: \datename, \subjclassname, and \proofname (in the amsthm package).

The \abstractname command is used in the article, report, amsart, and amsproc document classes. The letter document class uses a number of commands special to it: \pagename, \enclname, \ccname, and \headtoname.

If your document has photographs rather than figures, redefine \figurename:

```
\renewcommand{\figurename}{Photograph}
```

If you use the book or amsbook document class, you can even have

```
\renewcommand{\listfigurename}{List of Photographs}
```

9.1.6   Showing the meaning of commands

If you are defining a new command with \newcommand, and an error message advises that the command name is already in use, then it's useful to find out the meaning of the command. For instance, the \vct command is defined in sampart2.tex

\subjclassname	1991 Mathematics Subject Classification
\abstractname	Abstract
\alsoname	Also
\alsoseename	Also see
\appendixname	Appendix
\bibname	Bibliography
\ccname	Cc
\chaptername	Chapter
\contentsname	Contents
\datename	Date
\enclname	Enclosure
\figurename	Figure
\indexname	Index
\keywordsname	Key words and phrases
\listfigurename	List of Figures
\listtablename	List of Tables
\notesname	Notes
\headpagename	Page
\pagename	Page
\partname	Part
\proofname	Proof
\refname	References
\tablename	Table
\prefacename	Preface
\seename	See
\subjectname	Subject
\tablename	Table
\tocname	Table of Contents
\contentsname	Table of Contents
\headtoname	To

Table 9.1: Table of redefinable names in LaTeX

(in the ftp directory and in Appendix D). It would have been natural to call this new command \vec. If you do, you get the error message

```
! LaTeX Error: Command \vec already defined.
```

Here is how you can find out the definition of the \vec command.

   Get into interactive mode (see section 1.11.2) and type:

```
*\show \vec
```

LaTeX responds:

```
> \vec=macro:
```

```
->\mathaccent "017E .
<*> \show \vec
```

informing you that \vec is a command, and in fact a math accent (see section 4.9 and Appendix A). Now try \hangafter (see section 2.7.2):

```
*\show \hangafter
```

```
> \hangafter=\hangafter.
<*> \show \hangafter
```

The response indicates that \hangafter is a primitive command, defined in TeX itself. Redefining a primitive command is not a good idea.

Try one more command, \medskip (see section 2.8.2), to find out how big it is:

```
*\show \medskip
> \medskip=macro:
->\vspace \medskipamount
```

This indicates that the length is stored in \medskipamount. So use \show to ask what \medskipamount is:

```
*\show \medskipamount
> \medskipamount=\skip14.
```

which does not give much information. In fact, \medskipamount is different from the commands you have seen so far; it's a *length command*, containing the value of \medskip (see section 9.3.2). Ask for the value of a length command (or parameter) with the \showthe command:

```
*\showthe \medskipamount
```

```
> 6.0pt plus 2.0pt minus 2.0pt.
```

So \medskip is a vertical space of 6 points, which allows stretching or shrinking by up to 2 points.

LaTeX has many registers containing numbers, and parameters containing integers (such as 3), dimensions (such as 10.2pt), or lengths written in the form 6.0pt plus 2.0pt minus 2.0pt (called *glues* or *rubber length*, see section 9.3.2 and section E.2.2). Use \showthe for all these.

Alternatively, rather than going into interactive mode, type the \show and \showthe commands into your document. The explanation will appear on the monitor (and be written in the log file).

9.2   *User-defined environments*

Most user-defined commands are new commands. *User-defined environments*, as a rule, modify an existing environment. For instance, if you are not comfortable with the name of the `proof` environment (defined in the amsthm package) and would prefer the name demo, define

```
\newenvironment{demo}
   {\begin{proof}}
   {\end{proof}}
```

Note that this does not change the `proof` environment.

To modify an existing environment, type

```
\newenvironment{name}
   {begin_text}
   {end_text}
```

where *begin_text* contains the command \begin{*oldname*} and *end_text* contains the command \end{*oldname*}, where *oldname* is the name of the modified environment. For instance, the command,

```
\newenvironment{demo}
   {\begin{proof}\em}
   {\end{proof}}
```

defines the demo environment that typesets the proof emphasized. Note that the scope of \em is just the demo environment.

If an error shows up in such a user-defined environment, the message refers to the environment that was modified. For instance, if you misspell proof as prof when you define

```
\newenvironment{demo}
   {\begin{prof}\em}
   {\end{proof}}
```

then at the first use of demo you get the message:

```
! LaTeX Error: Environment prof undefined.
```

```
l.13 \begin{demo}
```

Or if you define:

```
\newenvironment{demo}
   {\begin{proof}\em}
   {\end{prof}}
```

at the first use of demo you get the message:

```
! LaTeX Error: \begin{proof} on input line 5 ended by \end{prof}.
```

1.14 \end{demo}

A \newenvironment can have arguments, but they can only be used in the *begin_text*. Here is a simple example. Define a theorem proclamation in the preamble (section 3.4), and then define

```
\newenvironment{theoremRef}[1]
    {\begin{theorem}\label{T:#1}}
    {\end{theorem}}
```

which is invoked with

```
\begin{theoremRef}{ label }
```

The theoremRef environment is a modification: it's a theorem that can be referenced (with the \ref command, of course); it invokes the theorem environment and defines T: *label* to be the label for cross-referencing.

Some environments (for instance, the equation environment in the presence of the amsmath package) can't be easily modified.

Tip   Do not give a new environment the name of a command.

For instance, if you define

```
\newenvironment{parbox}
    {...}
    {...}
```

then you get the error message:

```
! LaTeX Error: Command \parbox already defined.
```

Recall that a newly defined command remains effective only within its scope (see section 2.3.2). Now suppose you want to make a change, say redefining a counter, in the context of a few paragraphs. Of course, you can place braces around these paragraphs or define

```
\newenvironment{exception}
    {\relax}
    {\relax}
```

and then

```
\begin{exception}
   new commands
   body
\end{exception}
```

stands out better than a pair of braces. The \relax command "does nothing"; it is customary to include it in the definition to make it more readable.

Here is another example of a user-defined environment that is not a modification of an existing environment:

```
\newenvironment{vcenterpage}
               {\newpage\vspace*{\fill}}
               {\vspace*{\fill}\par\pagebreak}
```

This environment centers its body vertically on a new page.

The first argument of an environment may be *optional with a default value.* For instance,

```
\newenvironment{narrow}[1][3in]
   {\noindent\begin{minipage}{#1}}
   {\end{minipage}}
```

creates a narrow environment; it sets the body of the environment, not indented, wrapped in a column of width 3 inches:

```
\begin{narrow}
   This text was typeset in a \textttt{narrow} environment, not
   indented, wrapped in a column 3 inches wide.
\end{narrow}
```

which prints:

> This text was typeset in a narrow environment, not indented, wrapped in a column 3 inches wide.

However, you can give also an optional argument, for instance, 3.5in:

```
\begin{narrow}[3.5in]
   This text was typeset in a \textttt{narrow} environment, not
   indented, wrapped in a column 3 inches wide.
\end{narrow}
```

which prints the following (false) statement:

> This text was typeset in a narrow environment, not indented, wrapped in a column 3 inches wide.

See section 9.6.3 for another example of a user-defined environment.

Redefine an existing environment with the \renewenvironment command, which is similar to the \renewcommand command (see section 9.1.3).

9.2.1   *Short arguments*

You saw five commands that define new commands and environments; they define commands and environments that take arguments any number of paragraphs long. The *-ed versions of these commands define *short* commands and environments (see section 2.3.3) that take only one paragraph of text as an argument. (Note that for environments, the *-ed version makes the argument, not the body, short.) For instance,

```
\newcommand{\LB}[1]{{\large\bfseries#1}}
```

makes its argument large and bold. So

```
\LB{First paragraph.

Second paragraph.}
```

prints

⌐

 First paragraph.
 Second paragraph.

└

as expected. On the other hand, if you define

```
\newcommand*{\LB}[1]{{\large\bfseries#1}}
```

and then attempt to typeset the above example you get the error message:

```
! Paragraph ended before \LB was complete.
<to be read again>
                        \par
```

Short commands are preferable because of improved error checking.

9.3   *Numbering and measuring*

LATEX stores integers in *counters*; for instance, the section counter contains the current section number. LATEX also stores distance measurements in *length commands*; for instance, the \textwidth command contains the width of the text (345.0 points for this book).

In this section, you'll take a closer look at counters and length commands.

equation	footnote	figure	page	table
part	chapter	section	subsection	subsubsection
paragraph	subparagraph			
enumi	enumii	enumiii	enumiv	

Table 9.2: Standard LaTeX counters

9.3.1 Counters

Counters may be defined by LaTeX, by packages, or by the user.

Standard LaTeX counters

LaTeX automatically generates numbers for equations, sections, theorems, and so on. Each such number is associated with a *counter*. Table 9.2 shows the standard LaTeX counters. Most are self-explanatory; the fourth line implies that you can use four levels of mixed nested enumerated environments. In addition, for every proclamation *name*, there is a counter called name (see section 3.4).

Setting counters

The command for setting a counter is \setcounter. A number is generated by LaTeX by incrementing the appropriate counter, so if you want the current chapter to be numbered 3, set the chapter counter to 2 by typing

```
\setcounter{chapter}{2}
```

before the \chapter line. The only exception is the page number, which is first used to number the current page, and then incremented. So if you want to set the current page number to 63, then include the command

```
\setcounter{page}{63}
```

in the page.

LaTeX's standard counters are initialized and incremented by LaTeX. Sometimes you may want to manipulate them yourself. For example, suppose you are working on a book with a number of chapters. The main document book.tex contains the lines:

```
\include{intro}
\include{ch1}
\include{ch2}
\include{ch3}
. . .
```

When working on Chapter 3, add the line

```
\includeonly{ch3}
```

as explained in section 2.11.1. This will process Chapter 3 only; however, any time the book is typeset, all the aux files will be read, and at the end, written out. An alternate strategy is to have a book3.tex file containing the lines

```
\setcounter{chapter}{2}
\include{ch3}
```

and when book3.tex is typeset, the chapter will be properly numbered. You can also set

```
\setcounter{page}{63}
```

if the first page of this chapter is supposed to be 63.

There are disadvantages to this method: you lose all cross-references to other chapters, citations, and so on; and it's error-prone, since you may inadvertently leave in some of the \setcounter commands introduced for special processing. The advantage is increased processing speed, since LaTeX does not have to read all the aux files.

This example serves more as an illustration of how you can manipulate LaTeX's counters than as a practical suggestion.

Tip   If you need to manipulate LaTeX's counters, always look for solutions in which LaTeX will do the work for you.

Defining new counters

You can define your own counters. For example,

```
\newcounter{numb}
```

makes numb a new counter. In the definition, you can use an optional argument, which is the name of another counter:

```
\newcounter{numb}[hlnumb]
```

which will automatically reset numb to 0 whenever hlnumb changes value. This is the command LaTeX uses internally to number theorems within sections, for instance.

Rule ■ New counters
New counters should be defined in the preamble of the document; they can't be defined in a file read in with an \include command (see section 2.11.1).

If you violate this rule, there is no error message. However, when an \includeonly command does not include the file in which the counter is defined, you'll get a variety of confusing messages even if the counter is not used.

Counter styles

The visual representation of the counter numb can be displayed in the typeset document with

\thenumb

If you want to change the appearance of numb, issue the command

\renewcommand{\thenumb}{*new_format*}

where *new_format* specifies the new format.

A counter can be displayed in one of five styles:

arabic \arabic{*counter*} (1, 2, ...)

lowercase roman \roman{*counter*} (i, ii, ...)

uppercase roman \Roman{*counter*} (I, II, ...)

lowercase letters \alph{*counter*} (a, b, ... , z)

uppercase letters \Alph{*counter*} (A, B, ... , Z)

The default is arabic. Here is an example:

```
\renewcommand{\thechapter}{\arabic{chapter}}
\renewcommand{\thesection}{\thechapter-\arabic{section}}
\renewcommand{\thesubsection}%
    {\thechapter-\arabic{section}.\arabic{subsection}}
```

With these definitions, subsection 2 of section 1 of chapter 3 would be numbered in the form: 3-1.2.

To set the page numbering to lowercase roman in the introduction, and arabic in the rest of the book, your document should contain the lines:

```
\pagenumbering{roman}
\maketitle
\tableofcontents
\listoftables
\include{intro}
\pagenumbering{arabic}
\include{ch1}
\include{ch2}
...
```

In fact,

```
\pagenumbering{roman}
```

is defined in LaTeX as

```
\renewcommand{\thepage}{\roman{page}}
\setcounter{page}{0}
```

Here is another example using the amsmath package. The subequations environment (see section 4.15) uses the counter parentequation. To change the format of the equation numbers from (2a), (2b), and so on, to (2i), (2ii), and so on, in the subequations environment define:

```
\renewcommand{\theequation}{\theparentequation\roman{equation}}
```

Or if you prefer the format (2.i), (2.ii), and so on, define

```
\renewcommand{\theequation}{\theparentequation.\roman{equation}}
```

Counter arithmetic

The \stepcounter{hlnumb} command increments the counter hlnumb and sets to 0 all the counters that were defined with optional argument hlnumb. The variant \refstepcounter{hlnumb} does the same, and in addition sets the meaning of the \label. So after the above command, the next \label will refer to the value of the counter hlnumb.

You can do a little more arithmetic with

```
\addtocounter{counter}{n}
```

where n is an integer; for instance,

```
\setcounter{numb}{5}
\addtocounter{numb}{2}
```

will set numb to 7.

The value stored in a counter is made available by the \value command. This is mostly used with the \setcounter or \addtocounter commands. For instance, you can set one counter numb equal to the value of another counter oldnumb by typing

```
\setcounter{numb}{\value{oldnumb}}
```

Here is a typical example of counter manipulation. As is customary, you may want a theorem (in a theorem environment) followed by some corollaries (each in a corollary environment) always starting with Corollary 1. In other words, Theorem 1 should be followed by Corollary 1, Corollary 2, and so forth. Next comes Theorem 2. Normally, LaTeX would number the next corollary Corollary 3. To avoid this, one could precede the corollary with the command

```
\setcounter{corollary}{0}
```

but such a manual process is error-prone, at best.

Now accept the admonition of the tip on page 284, and let's see how you can get LaTeX do the work for you. In the preamble, type the proclamations:

```
\newtheorem{theorem}{Theorem}
\newtheorem{corollary}{Corollary}[theorem]
```

This almost works: Theorem 1 is followed by Corollary 1.1 and Corollary 1.2; while Theorem 2, followed by Corollary 2.1. Now redefine \thecorollary (see section 9.3.1):

```
\renewcommand{\thecorollary}{\arabic{corollary}}
```

so that Theorem 1 is followed by Corollary 1 and Corollary 2; while Theorem 2 is also followed by Corollary 1.

If you need to perform more complicated arithmetic with counters, use the calc package by Kresten K. Thorup and Frank Jensen (see section A.4 in *The LaTeX Companion*).

Two special counters

The secnumdepth and tocdepth counters control which sectional units are numbered and listed in the table of contents, respectively. For instance,

```
\setcounter{secnumdepth}{2}
```

sets secnumdepth to 2. As a result, chapters (if they are present in the document class), sections, and subsections are numbered, but subsubsections are not. This command must be placed in the preamble of the document.

9.3.2 Length commands

While a counter contains integers, a length command contains a *real number* and a *dimensional unit*.

Dimensional units

LaTeX recognizes seven dimensional units; five *absolute* units: cm centimeter, in inch, pc pica (1pc = 12pt), pt point (1in = 72.27pt), and mm millimeter; and two *relative* units: em, which is about the width of the letter M in the current font, and ex, which is about the height of the letter x in the current font.

LaTeX defines a very large number of length commands. For instance, the *The LaTeX Companion* in section 4.1 lists 17 length commands for page layout alone (you can find some of them in Figure 6.4); a list environment sets about a dozen length commands (see Figure 9.1). Length commands are defined for almost every aspect of LaTeX's work, including displayed math environments; a complete list would probably contain a few hundred. A good number of these are listed in Lamport [30] and *The LaTeX Companion*; many more are hidden in packages such as amsmath.

The most common length commands are \parindent (the amount of indentation at the beginning of a paragraph), \parskip (the extra vertical space inserted between paragraphs—normally set to zero), and \textwidth (the width of text on a page). A more esoteric example is \marginparpush (the minimal vertical space between two marginal notes). Luckily, you do not have to be familiar with many length commands, since LaTeX and the document class set them for you.

Setting length

The \setlength command sets (or resets) the value of a length command. So

\setlength{\textwidth}{3in}

will make a very narrow page. The first argument of \setlength is a length command, *not the command name*; that is,

\setlength{textwidth}{3in} % Bad

is incorrect. The second argument of \setlength is a real number with a dimensional unit, for instance, 3in, and *not a real number alone*; in other words,

\setlength{\textwidth}{3} % Bad

is also incorrect.

Tip   A common mistake is to write something like

\setlength{\marginpar}{0}

Instead, write

\setlength{\marginpar}{0pt}

or

\setlength{\marginpar}{0in}

or whatever. Just be sure to include a dimensional unit.

The \addtolength command adds a quantity to the value of a length command. For instance,

\addtolength{\textwidth}{-10pt}

narrows the page width by 10 points.

When LaTeX typesets some text (or math), it creates a "box". Three measurements are used to describe the size of the box: the width, the height (from the baseline to the top), and the depth (from the baseline to the bottom). For instance, the box "aa" has width 10.00003pt, height 4.30554pt, and depth 0pt. The box "ag" has the same width and height, but it has depth 1.94444pt. The box "Ag" has width 12.50003pt, height 6.83331pt, and depth 1.94444pt. The commands

```
\settowidth
\settoheight
\settodepth
```

each have two arguments; the first argument is a length command, while the second is text (or math) to be typeset by LaTeX. The width of the box corresponding to the second argument is assigned to the length command. It should be clear now how \phantom can be defined with \hspace and \settowidth.

To perform more complicated arithmetic with length commands, use the calc package of Kresten K. Thorup and Frank Jensen (see section A.4 in *The LaTeX Companion*).

Rubber lengths

In addition to rigid lengths such as 3in, LaTeX can also set a *rubber length*, that is, a length that is allowed to stretch and shrink. Here is an example:

```
\setlength{\stretchspace}{3in plus 10pt minus 8pt}
```

Assuming \stretchspace is a length command, this assigns the value 3 inches, which can stretch by 10 points, or shrink by 8 points, as the case may be. So if you make a box of this width, it'll be 3 inches wide, plus up to 10 points, or minus up to 8 points.

Stretchable vertical spaces are used, for instance, before and after displayed text environments; the spaces are adjusted to make the page look good. An example will be found in section 9.1.6: \medskipamount is defined as

```
6.0pt plus 2.0pt minus 2.0pt
```

See section 9.6.3 for more examples.

Defining new length commands

A new length command is defined with the \newlength command. So

```
\newlength{\mylength}
```

makes \mylength a length command. Now type

```
\setlength{\mylength}{3in}
```

and make a \parbox of width \mylength:

\parbox{\mylength}{...}

This technique has editorial advantages: you can uniformly change the width of many \parboxes in the document with a single \setlength command. It also allows relative size changes. For example, define

\newlength{\shorterlength}
\setlength{\shorterlength}{\mylength}
\addtolength{\shorterlength}{-.5in}

and then \parbox{\shorterlength}{...} will always typeset in a column half an inch narrower than the other parboxes.

9.4   *Delimited commands*

Delimited commands provide a way of writing the source document in a more readable fashion.

They have to be defined with (TeX's) \def command (and not with LaTeX's \newcommand). To define such a command, write \def, and the new command (not in braces); the definition then follows in braces. For instance, the first command defined in section 9.1.1:

\newcommand{\la}{\leftarrow}

may be typed as follows

\def\la{\leftarrow}

LaTeX does not check whether the new command name is already in use, so \def is different from \newcommand, \renewcommand, or \providecommand (see section 9.1.3). If the \la command was previously defined, the original definition is overridden.

Tip   The responsibility is *yours* when redefining a command with \def; LaTeX does not provide any protection.

In Example 3 of section 9.1.2, a command with three arguments was introduced for typing congruences (using the amsmath package):

\newcommand{\con}[3]{#1\equiv#2\pod{#3}}

Then $\con{a}{b}{\theta}$ prints: $a \equiv b \ (\theta)$. This saves a little typing, but it does not make the source file much easier to read; to achieve this, use *delimited commands*.

Let's start with a simple example, by defining a command for vectors:

```
\def\vv<#1>{\langle#1\rangle}
```

Note that \vv is a command with one argument, #1. When invoked, it'll print the delimiter ⟨, the argument, and then the delimiter ⟩.

Note that in the definition of \vv, the argument #1 is delimited by < and >. When the command is invoked, *the argument must be delimited the same way.*

So for the vector $\langle a, b \rangle$ invoke \vv with

```
\vv<a,b>
```

which looks somewhat like a vector, with \vv there as a reminder.

You have to be careful with delimited commands, because the math space rule (see section 4.2) does not hold, neither in the definition, nor in the invocation. So if in the definition

```
\def\vv<␣#1>{\langle#1\rangle}
```

there is a space before #1, then $\vv<a,b>$ gives the error message:

```
! Use of \vv doesn't match its definition.
l.12 $\vv<a
          ,b>$
```

which is clear enough. Now if the space is on the other side:

```
\def\vv<#1 >{\langle#1\rangle}
```

the error message is:

```
Runaway argument?
a,b>$
! Paragraph ended before \vv was complete.
<to be read again>
                    \par
```

The moral is that if you use delimited commands, be very careful that the invocation completely matches the definition.

Returning to the congruence example (using the amsmath package), you may write:

```
\def\con#1=#2(#3){#1\equiv#2\pod{#3}}
```

so that $\con a=b(\theta)$ prints: $a \equiv b \ (\theta)$. In the source, \con a=b(\theta) looks a bit like a congruence and is easier to read.

There is only one catch. Suppose you want to typeset the formula

$$x = a \equiv b \quad (\theta)$$

If you type `$\con x=a=b(\theta)$`, LATEX will print $x \equiv a = b \ (\theta)$. Indeed, "x" is delimited on the right by the first =; hence, the first argument is x. The second argument is delimited by the first = and the left parenthesis; hence, the second argument is a=b. This is not what was intended, however. In such cases, help LATEX find the correct first argument by typing

`$\con{x=a}=b(\theta)$`

In section 2.3.1, we discussed the problem of typing a command such as `\TeX` (the example there was `\today`) in the form: `\TeX\␣` so that "TEX" will be typeset as a separate word. The problem is that if you just type `\TeX` without the trailing `\␣`, then TEX is merged with the next word, and there is no error message. One solution is to use a delimited command:

`\def\tex/{\TeX}`

Now to get TEX, type `\tex/`; if a space is needed after it, then type `\tex/␣`. If you forget the closing /, *you get an error message.*

A better solution to this problem is the use of the xspace package (see section 9.1.1); however, in many documents (including the $\mathcal{AMS}$ documentation) you will find the delimited construct, so you should be familiar with it.

9.5   *A custom command file*

User-defined commands, of course, are a matter of individual need and taste. I have collected in the `lattice.sty` file (which you can find in the `ftp` directory—see page 4) some commands I use in writing papers in lattice theory. I hope that this model will help you develop a command file of your own. Please, remember that everything you see in this section is a reflection of *my* work habits. Many experts disagree with one or another aspects of what is done in this section; use whatever suits your needs.

This file was named `lattice.sty` and not `lattice.tex` so that it can be invoked with `\usepackage` instead of `\input`. This has a number of advantages, as you'll see.

Command names should be *mnemonic*; if you can't easily remember a name, rename it. This implies that your command file should not be too large unless you have an unusual ability to recall abbreviations.

Here are the first few lines of the command file `lattice.sty`:

```
\NeedsTeXFormat{LaTeX2e}
\ProvidesPackage{lattice}
   [1995/03/15  Commands for lattices, version 5.2]
\RequirePackage{amsmath}
\RequirePackage{amssymb}
\RequirePackage{eucal}
```

The first line \NeedsTeXFormat{LaTeX2e} will give an error message if this file is typeset with LaTeX 2.09. The second and third lines provide information that will be written in your log file. The next three lines declare what packages are required. If they have already been loaded, these lines will be ignored; otherwise, the missing packages will be loaded. A package invoked with \RequirePackage is not loaded twice.

If you start your article with

```
\documentclass{article}
\usepackage{lattice}
```

then the \listfiles command (see section 1.11.4) produces the following list:

```
article.cls    1994/12/09 v1.2x Standard LaTeX document class
size10.clo     1994/12/09 v1.2x Standard LaTeX file (size option)
lattice.sty    1995/03/15 Commands for lattices, version 5.2
amsmath.sty    1995/02/23 v1.2b AMS math features
amstext.sty    1995/01/25 v1.2
amsgen.sty     1995/01/31 v1.2a
amsbsy.sty     1995/01/20 v1.2
amsopn.sty     1995/02/20 v1.2a operator names
amssymb.sty    1995/02/01 v2.2a
amsfonts.sty   1995/02/01 v2.2b
eucal.sty      1994/12/13 v2.1e LaTeX package Euler Script
```

Next we have in lattice.sty the commands for lattices and sets:

```
% Lattice operations
\newcommand{\jj}{\vee}% join
\newcommand{\mm}{\wedge}% meet
\newcommand{\JJ}{\bigvee}% big join
\newcommand{\MM}{\bigwedge}% big meet

% Set operations
\newcommand{\uu}{\cup}% union
\newcommand{\ii}{\cap}% intersection
\newcommand{\UU}{\bigcup}% big union
\newcommand{\II}{\bigcap}% big intersection

% Sets
\newcommand{\ci}{\subseteq}% contained in with equality
\newcommand{\nc}{\nsubseteq}% not \ci
\newcommand{\nci}{\nc}% an alias for \nc
\newcommand{\ce}{\supseteq}% containing with equality
\newcommand{\nce}{\nsupseteq}% not \ce
```

```
\newcommand{\nin}{\notin}% not \in
\newcommand{\es}{\varnothing}% the empty set
\newcommand{\set}[1]{\{#1\}}% set
\newcommand{\setm}[2]{\{\,#1\mid#2\,\}}% set with a middle
\def\vv<#1>{\langle#1\rangle}% vector


% Partial ordering
\newcommand{\nle}{\nleq}% not \leq
```

So $a \jj b$ prints $a \vee b$ and $A \ci B$ prints $A \subseteq B$, and so on. Note that the original commands are not redefined; if a coauthor prefers $a \vee b$ in lieu of $a \jj b$, it's still available.

With the \set command, type the set $\{a, b\}$ as $\set{a,b}$, which is easier to read. Similarly, type the vector $\langle a, b \rangle$ as $\vv<a,b>$, so it looks like a vector.

Next I map the Greek letters to the keyboard. For some Greek letters, I prefer to use the variants, a matter of individual taste. (It is also a matter of taste *whether* to map the Greek letters to the keyboard at all, and how far one should go in abbreviating common commands.)

```
% Greek letters
\newcommand{\ga}{\alpha}
\newcommand{\gb}{\beta}
\newcommand{\gc}{\chi}
\newcommand{\gd}{\delta}
\renewcommand{\ge}{\varepsilon}% use \geq for >=
\newcommand{\gf}{\varphi}
\renewcommand{\gg}{\gamma}% old use >>
\newcommand{\gh}{\eta}
\newcommand{\gi}{\iota}
\newcommand{\gj}{\theta}
\newcommand{\gk}{\kappa}
\newcommand{\gl}{\lambda}
\newcommand{\gm}{\mu}
\newcommand{\gn}{\nu}
\newcommand{\go}{\omega}
\newcommand{\gp}{\pi}
\newcommand{\gq}{\theta}
\newcommand{\gr}{\varrho}
\newcommand{\gs}{\sigma}
\newcommand{\gt}{\tau}
\newcommand{\gu}{\upsilon}
\newcommand{\gv}{\vartheta}
\newcommand{\gw}{\omega}
```

```
\newcommand{\gx}{\xi}
\newcommand{\gy}{\psi}
\newcommand{\gz}{\zeta}

\newcommand{\gC}{\Xi}
\newcommand{\gG}{\Gamma}
\newcommand{\gD}{\Delta}
\newcommand{\gF}{\Phi}
\newcommand{\gL}{\Lambda}
\newcommand{\gO}{\Omega}
\newcommand{\gP}{\Pi}
\newcommand{\gQ}{\Theta}
\newcommand{\gS}{\Sigma}
\newcommand{\gU}{\Upsilon}
\newcommand{\gW}{\Omega}
\newcommand{\gX}{\Xi}
\newcommand{\gY}{\Psi}
```

I also introduce shorter command names for text and math font commands by abbreviating "text" to "t" (so \textbf becomes \tbf) and "math" to "m" (so \mathbf becomes \mbf).

```
% Font commands
\newcommand{\tbf}{\textbf}% text bold
\newcommand{\tit}{\textit}% text italic
\newcommand{\tsl}{\textsl}% text slanted
\newcommand{\tsc}{\textsc}% text small caps
\newcommand{\ttt}{\texttt}% text typewriter
\newcommand{\trm}{\textrm}% text roman
\newcommand{\tsf}{\textsf}% text sans serif
\newcommand{\tup}{\textup}% text upright

\newcommand{\mbf}{\mathbf}% math bold
\renewcommand{\mit}{\mathit}% math italic
\newcommand{\msf}{\mathsf}% math sans serif
\newcommand{\mrm}{\mathrm}% math roman
\newcommand{\mtt}{\mathtt}% math typewriter

\newcommand{\B}{\boldsymbol}
   % Bold math symbol
\DeclareMathAlphabet{\Bi}{OT1}{cmm}{b}{it}
   % Bold math italic
\newcommand{\C}{\mathcal}
```

```
   % Euler Script - caps only
\newcommand{\D}{\mathbb}
   % Doubled - blackboard bold - caps only
\newcommand{\E}{\mathcal}% same as \C
   % Euler Script - caps only
\newcommand{\F}{\mathfrak}% Fraktur
```

The math alphabets are invoked with commands with arguments: \B for **Bold**, \Bi for **bold** math *italic*, \C for Calligraphic, \D for blackboard bold (**Double**), and \F for Fraktur (German Gothic) (see section 4.14.2). Since I invoke the eucal package with no option, Euler Script substitutes for calligraphic (see section 4.14.1). Notice that I retained both \C (calligraphic) and \E (Euler Script) for the Euler Script alphabet.

Here are some other commands:

```
% Miscellaneous
\newcommand{\nl}{\newline}
\newcommand{\ol}{\overline}
\newcommand{\ul}{\underline}
\def\con#1=#2(#3){#1\equiv#2\pod{#3}}
   % congruence: \con a=b(\gQ)
\providecommand{\bysame}{\makebox[3em]{\hrulefill}\thinspace}
\newcommand{\q}{\quad}
\newcommand{\qq}{\qquad}
```

```
% Referencing sections and declarations
\newcommand{\refS}[1]{section~\ref{S:#1}}
\newcommand{\refSS}[1]{section~\ref{SS:#1}}
\newcommand{\refT}[1]{Theorem~\ref{T:#1}}
\newcommand{\refL}[1]{Lemma~\ref{L:#1}}
\newcommand{\refD}[1]{Definition~\ref{D:#1}}
\newcommand{\refC}[1]{Corollary~\ref{C:#1}}
```

```
\endinput
```

Sections and declarations are referenced with \ref*X*{*label*}, where *X* is S for sections, SS for subsections, T for theorem, L for lemma, D for definition, and C for corollary.

In Appendix C (also in the ftp directory), we rewrite the sampart.tex sample article utilizing the user-defined commands of this section.

9.6 *Custom lists*

Although LaTeX provides three types of list environments (see section 3.1), it's often necessary to create one of your own. The `list` environment will then do the job. In fact, the `list` environment is used by LaTeX to define many of the standard environments:

- the three list environments (section 3.1);
- the quote, quotation, verse environments (section 3.3);
- the proclamations (section 3.4);
- the style environments, center, flushleft, flushright (section 3.8);
- the thebibliography environment (section 6.5.1).

9.6.1 *Length commands for the* list *environment*

The general layout for a list is shown in Figure 9.1: it consists of six horizontal measurements and three vertical measurements which LaTeX figures out from six length commands for the horizontal measurements and from four length commands for the vertical measurements. I now list these length commands:

Vertical length commands

\topsep is the vertical space between the first item and the preceding text, and also between the last item and the following text. Actually, this is not really true; this separating space is of height equal to the sum of \topsep and \parskip (the extra vertical space inserted between paragraphs—normally set to zero). Optionally, you can increase this space by adding \partopsep provided the list environment starts a new paragraph.

\parsep is the space between paragraphs of the same item.

\itemsep added to \parsep is the space between items. In other words, the space between items is the sum of \itemsep and \parsep.

All the vertical length commands are rubber lengths (see section 9.3.2).

Horizontal length commands

The margin for the `list` environment is the margin of the surrounding text. In case of normal text, this is the normal margin. If the list is, say, nested within a list, the margin of the surrounding text is, as a rule, narrower than normal.

The \leftmargin and \rightmargin length commands measure the distance of the item box from the left and right margin respectively.

The label is the text given in the (optional) argument of the \item command or is provided as a default; it is typeset in a box of width \labelwidth, which is placed \itemindent units from the left margin, and it is separated by a space of

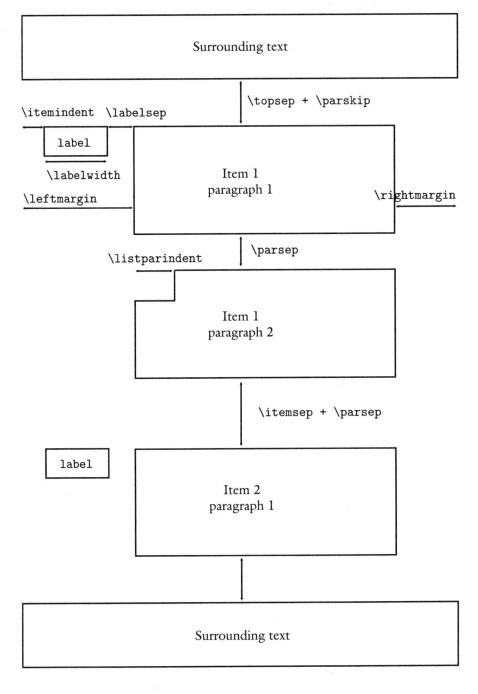

Figure 9.1: The layout of a custom list

\labelsep units from the text box. If the label is too wide, it's typeset in its natural width, so the first line in the text box will be indented.

The second and subsequent paragraphs of an item will be typeset indented by \listparindent units.

9.6.2 *The* list *environment*

One creates a custom list with the list environment, which is invoked in the following form:

```
\begin{list}{default_label}{declarations}
\item item1
\item item2
...
\end{list}
```

where *default_label* sets the label for those items that do not specify it (similar to the optional argument of the \item command), and *declarations* sets the vertical and horizontal length commands (and whatever else is required) for the list.

Here is a very simple example:

Here are the most important LaTeX rules about spaces in text, sentences, and paragraphs:

◇ **Rule 1:** Two or more spaces in text are the same as one.

◇ **Rule 2:** A blank line (that is, two end-of-line characters separated only by blanks and tabs) indicates the end of a paragraph.

Rules 1 and 2 make typing and copying very convenient.

I use ◇ ($\diamondsuit$) as a default label, and I set the item box 0.5 inches from either margin. So this example is typed as follows:

```
Here are the most important \LaTeX\ rules about spaces in text,
sentences, and paragraphs:
\begin{list}{$\diamondsuit$}{\setlength{\leftmargin}{.5in}
                            \setlength{\rightmargin}{.5in}}
\item \textbf{Rule 1:} Two or more spaces in text are
the same as one.
\item \textbf{Rule 2:} A blank line (that is, two end-of-line
characters separated only by blanks and tabs) indicates the end
of a paragraph.
```

```
\end{list}
Rules~1 and 2 make typing and copying very convenient.
```

Here is a second variant:

⌐

Here are the most important LᴬTEX rules about spaces in text, sentences, and paragraphs:

Rule 1: Two or more spaces in text are the same as one.

Rule 2: A blank line (that is, two end-of-line characters separated only by blanks and tabs) indicates the end of a paragraph.

Rules 1 and 2 make typing and copying very convenient.
└

In this example, I dropped the optional *default_label* and typed **Rule 1:** and **Rule 2:** as (optional) arguments of the \item commands:

```
Here are the most important \LaTeX\ rules about spaces in text,
sentences, and paragraphs:
\begin{list}{}{\setlength{\leftmargin}{.5in}
               \setlength{\rightmargin}{.5in}}
\item[\textbf{Rule 1:}] Two or more spaces in text are
the same as one.
\item[\textbf{Rule 2:}] A blank line (that is, two end-of-line
characters separated only by blanks and tabs) indicates the end
of a paragraph.
\end{list}
Rules~1 and 2 make typing and copying very convenient.
```

For further simple examples, look in various document class files to see how standard environments such as verse, quote, and so on, are defined.

Using counters

It is not very LᴬTEX-like to provide the numbers for the rules in the examples above; it would be more logical for LᴬTEX to do the numbering. The following is a more LᴬTEX-like coding of the second example:

```
Here are the most important \LaTeX\ rules about spaces in text,
sentences, and paragraphs:
\newcounter{spacerule}
\begin{list}{\textbf{Rule \arabic{spacerule}:}}
            {\setlength{\leftmargin}{.5in}
             \setlength{\rightmargin}{.5in}
```

```
            \usecounter{spacerule}}
\item Two or more spaces in text are
the same as one.
\item A blank line (that is, two end-of-line
characters separated only by blanks and tabs) indicates the end
of a paragraph.
 \end{list}
Rules~1 and 2 make typing and copying very convenient.
```

Note that

1. I declared the counter (before the `list` environment) with the line

   `\newcounter{spacerule}`

2. I defined the *default_label* as

   `Rule \arabic{spacerule}:`

3. In the *declarations*, I declared that the counter `spacerule` be used with the clause

   `\usecounter{spacerule}`

9.6.3 Two complete examples

In the previous examples, I set the values of `\leftmargin` and `\rightmargin`; the other length commands were not redefined, so their values were those set by the document class. In the following examples, I'll set the length of many more length commands.

Example 1   This is what I would like to get:

Here are the most important LaTeX rules about spaces in text, sentences, and paragraphs:

Rule 1:   *Two or more spaces in text are the same as one.*

Rule 2:   *A blank line (that is, two end-of-line characters separated only by blanks and tabs) indicates the end of a paragraph.*

Rules 1 and 2 make typing and copying very convenient.

This is typed as follows:

```
Here are the most important \LaTeX\ rules about spaces in text,
sentences, and paragraphs:
\newcounter{spacerule}
\begin{list}{\upshape \bfseries Rule \arabic{spacerule}:}
           {\setlength{\leftmargin}{1.5in}
            \setlength{\rightmargin}{.6in}
            \setlength{\labelwidth}{1.0in}
            \setlength{\labelsep}{.2in}
            \setlength{\parsep}{0.5ex plus 0.2ex minus 0.1ex}
            \setlength{\itemsep}{0ex plus 0.2ex minus 0ex}
            \usecounter{spacerule}
            \itshape}
\item Two or more spaces in text are the same as one.
\item A blank line (that is, two end-of-line characters
separated only by blanks and tabs) indicates the end
of a paragraph.
\end{list}
Rules~1 and 2 make typing and copying very convenient.
```

Note the following:

1. I declared the counter as in the previous example.
2. The last item in *declarations* is \itshape, which typesets the entire list in italics.
3. The *default_label* is defined as

   ```
   \upshape \bfseries Rule \arabic{spacerule}:
   ```

   My first attempt was to define it as

   ```
   \bfseries Rule \arabic{spacerule}:
   ```

   which typeset "Rule" in bold italics. To prevent this, I start the *default_label* with \upshape.
4. The left margin is set to 1.5 inches and the right margin to 0.6 inches:

   ```
   \setlength{\leftmargin}{1.5in}
   \setlength{\rightmargin}{.6in}
   ```

5. Next I set the width of the label to 1 inch, and the label separation to 0.2 inches:

   ```
   \setlength{\labelwidth}{1.0in}
   \setlength{\labelsep}{.2in}
   ```

6. Finally, I set the paragraph separation to 0.5 ex (allowing stretching by 0.2 ex and shrinking by 0.1 ex) and the item separation to 0.5 ex (allowing stretching by 0.4 ex and shrinking by 0.1 ex) by

```
\setlength{\parsep}{0.5ex plus 0.2ex minus 0.1ex}
\setlength{\itemsep}{0ex plus 0.2ex minus 0ex}
```

The numbers for item separation are formed by adding up the numbers declared for \parsep and \itemsep.

A complicated list such as this should be defined as a new environment. For instance, you could define a myrules environment:

```
\newenvironment{myrules}
  {\begin{list}
    {\upshape \bfseries Rule \arabic{spacerule}:}
    {\setlength{\leftmargin}{1.5in}
    \setlength{\rightmargin}{.6in}
    \setlength{\labelwidth}{1.0in}
    \setlength{\labelsep}{.2in}
    \setlength{\parsep}{0.5ex plus 0.2ex minus 0.1ex}
    \setlength{\itemsep}{0ex plus 0.2ex minus 0ex}
    \usecounter{spacerule}
    \itshape} }
  {\end{list}}
```

and then

```
\begin{myrules}
  \item Two or more spaces in text are the same as one.
  \item A blank line (that is, two end-of-line
characters separated only by blanks and tabs) indicates the end
of a paragraph.
\end{myrules}
```

typesets the first example as above.

Example 2   In section 2.7.2, we discussed the formatting of the following type of glossary:

sentence is a group of words terminated by a period, exclamation point, or question mark.

paragraph is a group of sentences terminated by a blank line or by the \par command.

Now do this as a custom list:

```
\begin{list}{}
  {\setlength{\leftmargin}{30pt}
```

```
\setlength{\rightmargin}{0pt}
\setlength{\itemindent}{14pt}
\setlength{\labelwidth}{40pt}
\setlength{\labelsep}{5pt}
\setlength{\parsep}{0.5ex plus 0.2ex minus 0.1ex}
\setlength{\itemsep}{0ex plus 0.2ex minus 0ex}}
\item[\textbf{sentence}\hfill] is a group of words terminated
    by a period, exclamation point, or question mark.
\item[\textbf{paragraph}\hfill] is a group of sentences
    terminated by a blank line or by the \com{par} command.
\end{list}
```

There is nothing new in this except for the \hfill commands in the optional arguments to left adjust the labels; with the long words in the example this is not necessary, but with shorter ones it must be done.

See section 3.2.2 of *The LATEX Companion* for more complicated custom lists, and section 3.2.1 on how to customize the three standard list environments.

9.6.4   *The* trivlist *environment*

LATEX provides a trivlist environment, more for programmers than for users. The environment is invoked in the form

```
\begin{trivlist}
    body
\end{trivlist}
```

It is similar to the list environment except that there are no arguments, and all the length commands are trivially set (most to 0 points, except for \listparindent and \parsep which are set equal to \parindent and \parskip, respectively). For instance, LATEX defines the center environment as follows:

```
\begin{trivlist}
    \centering \item[]
\end{trivlist}
```

9.7   *Custom formats*

At some point, you'll probably become annoyed at how long it takes LATEX to process the lines

```
\documentclass{article}
\usepackage{amssymb,amsmath}
```

You can speed this up with custom formats.

First, create a new copy of `latex.ltx`, call it `tmplatex.ltx`. The last line of `tmplatex.ltx` before \endinput is \dump. Comment it out.

Create a document `custart.tex` (custom article) that contains all the lines of your article up to the \begin{document} line, and add the line

```
\input tmplatex.ltx
```

to the beginning and the line \dump to the end. For instance, a very rudimentary `custart.tex` may read:

```
\input tmplatex.ltx
\documentclass{article}
\usepackage{amssymb,amsmath}
\dump
```

On the other hand, for `samplart2.tex` it would be a little more complex:

```
\input tmplatex.ltx
\documentclass{amsart}
\usepackage{lattice}
\usepackage[notcite]{showkeys}
\usepackage{xspace}

\theoremstyle{plain}
\newtheorem{theorem}{Theorem}
\newtheorem{corollary}{Corollary}
\newtheorem*{main}{Main~Theorem}
\newtheorem{lemma}{Lemma}
\newtheorem{proposition}{Proposition}

\theoremstyle{definition}
\newtheorem{definition}{Definition}

\theoremstyle{remark}
\newtheorem*{notation}{Notation}

\numberwithin{equation}{section}

\newcommand{\JJm}[2]{\JJ(\,#1\mid#2\,)}%
    % big join with middle used as: \JJm{a}{a < 2}
\newcommand{\Prodm}[2]{\gP(\,#1\mid#2\,)}
    % product with a middle
\newcommand{\Prodsm}[2]{\gP^{*}(\,#1\mid#2\,)}
    % product * with a middle
\newcommand{\vct}[2]{\vv<\dots,0,\dots,\overset{#1}{#2},%
```

```
\dots,0,\dots>}% special vector
\newcommand{\fp}{\ensuremath{\F{p}}\xspace}
   % Fraktur p in text or math
\newcommand{\Ds}{\ensuremath{D^{\langle2\rangle}}\xspace}
   % \Ds in text or math
\dump
```

Typeset `custart.tex` with `initex`. Name the resulting format file `art.fmt`. In your document, comment out all the lines that have been included in the document `custart.tex`, and add the line:

```
% Typeset with art format
```

Next time you typeset specifying this format, you'll be surprised how fast LaTeX gets to the body of the document. Before you submit the document for publication, *undo these changes.* The editor of the journal does not have the `art` format.

An alternative approach is to use David Carlisle's mylatex.ltx, which can be found on the CTAN (see section G.2 on how to get it).

Bibliography Index

PART V

Long bibliographies and indexes

BIBTEX

The BIBTEX program, written by Oren Patashnik, assists the LaTeX user in compiling longer bibliographies. Short bibliographies can easily be placed in the document itself (see section 6.5.1).

It takes a bit of an effort to learn BIBTEX. However in the long run, you'll find that the advantages of building bibliographic databases, which can be reused and shared, outweigh the disadvantage of a somewhat steep learning curve.

The bibliographic databases—the bib files—contain the bibliographic entries. First we'll discuss in section 10.1 the format of these entries. Then section 10.2 describes how to use BIBTEX to create bibliographies.

10.1 The database

To use BIBTEX for bibliographies, you first have to learn how to assemble the databases BIBTEX will access. The present section explains how to do that.

BIBTEX uses a "style file", called a "bibliographic style file", or bst file, to format an entry. To simplify our discussions, I use only one style file: the $\mathcal{AMS}$ plain style, amsplain.bst (which is part of the $\mathcal{AMS}$ distribution—see section G.2). All the examples are shown in this style, and several of the comments I make are true only of this style. If you choose to use a different bst file, look in the documentation for the special rules of that particular style.

10.1.1 *Entry types*

A bibliographic entry is given in "pieces" called *fields*. The bibliographic style (see section 10.2.2) specifies how the fields are typeset. Here are two typical entries:

```
@BOOK(gM68,
    author = "George A. Menuhin",
    title = "Universal Algebra",
    publisher = "D.~van Nostrand",
    address = "Princeton-Toronto-London-Melbourne",
    year = 1968,
    )
```

```
@ARTICLE(eM57,
    author = "Ernest T. Moynahan",
    title = "On a problem of {M}.~{H}. {S}tone",
    journal = "Acta Math. Acad. Sci. Hungar.",
    pages = "455--460",
    volume = 8,
    year = 1957,
    )
```

The first *entry type* BOOK is marked by @followed by the left delimiter (. The matching right delimiter) indicates the end of the entry.

The string "@BOOK(" is followed by the label gM68 designating the name of the entry; refer to this bibliographic entry in the document with \cite{gM68}. The label is followed by a comma and a series of fields. In this example, there are five fields: author, title, publisher, address, and year. A field starts with a field name, followed by "=" and the value of the field in double quotes ("); be sure to use " and *not* LATEX double quotes `` or ''. Numeric field values (consisting entirely of digits) do not need to be enclosed in double quotes (year in the above examples; volume in the second example, and number in some of the examples that follow). There must be a comma before each field (the comma before the first field was placed after the label).

There are many entry types:

ARTICLE an article in a journal or magazine;

BOOK a book with an author (or editor) and a publisher;

BOOKLET a printed work without a publisher;

INBOOK a part of a book, such as a chapter or a page range that, in general, is not titled or authored separately;

INCOLLECTION a part of a book with its own title (and perhaps author);

INPROCEEDINGS an article in a conference proceedings with its own title and perhaps author;

MANUAL technical documentation;

MASTERSTHESIS a Master's thesis;

MISC whatever does not fit in any other category;

PHDTHESIS a Ph.D. thesis;

PROCEEDINGS the proceedings of a conference;

TECHREPORT a report published by a school or institution;

UNPUBLISHED an unpublished paper.

Each entry has a number of *fields* chosen by the style from the following list:

address	author	booktitle
chapter	crossref	edition
editor	howpublished	institution
journal	key	language
month	note	number
organization	pages	publisher
school	series	title
type	volume	year

Note that the language field is specific to $\mathcal{AMS}$ bibliographic style files. Also you may add fields not recognized by the particular style file you use; such fields will be ignored.

Tip   BIBTEX does not care whether you use uppercase or lowercase letters for entry types and fields. In this book, the entry types are in uppercase and fields are in lowercase.

BIBTEX allows you to delimit the entry itself with braces instead of parentheses, and to delimit the value of the fields with braces instead of the double quotes.

Tip   It is optional to place a comma after the last field. It is recommended that you use a trailing comma so that if you append a field to the entry, the required comma will be there.

For each entry type there are *required fields* and *optional fields*. In subsequent examples in this section, two examples of each entry type are given. The first example of an entry type is a typical example, while the second example is a maximal one, showing all possible fields.

Tip   Make sure you type the field name correctly; if there is an error in the field name, BIBTEX will ignore the field. However, there is a warning message if a required field is missing.

All the examples in this section can be found in the `template.bib` file in the `ftp` directory—see page 4.

Rule ■ Uppercase in titles

The $\mathcal{AMS}$ bibliographic style file converts the title to lowercase, except for the first letter. If you need a letter in uppercase, put it in braces. The same rule applies for the `edition` field. (Most other style files—including the standard BIBTEX style files—do this conversion for titles of non-book-like entries only.)

In the second example above, three letters in the title are enclosed in braces so that they'll not be converted to lowercase.

Rule ■ Punctuating titles

You do not need to put a period at the end of a title; the bibliographic style file will supply it, if necessary.

Tip   BIBTEX and the bibliographic style file automatically handle for you a number of things that you have to handle yourself when typing text.

1. You do not have to mark periods for abbreviations (in the form `.\`⌴—see section 2.2.2) in the name of a journal, so

   ```
   journal = "Acta Math. Acad. Sci. Hungar.",
   ```

   will be typeset correctly.
2. You can type a single hyphen for a page range (instead of the usual `--`, see section 2.4.2) in the pages field; so

   ```
   pages = "455-460"
   ```

will produce an en-dash.

3. You do not have to type nonbreakable spaces (with ~, see section 2.4.3) in the author or editor fields:

```
author = "George A. Menuhin",
```

is correct (normally one would type George~A. Menuhin).

10.1.2   *Articles*

Entry type: ARTICLE
Required fields: author, title, journal, year, pages
Optional fields: volume, number, language, note

Examples:

13. Ernest T. Moynahan, *On a problem of M. H. Stone*, Acta Math. Acad. Sci. Hungar. **8** (1957), 455–460.
14. Ernest T. Moynahan, *On a problem of M. H. Stone*, Acta Math. Acad. Sci. Hungar. **8** (1957), no. 5, 455–460 (English), Russian translation available.

typed as

```
@ARTICLE(eM57,
    author = "Ernest T. Moynahan",
    title = "On a problem of {M}.~{H}. {S}tone",
    journal = "Acta Math. Acad. Sci. Hungar.",
    pages = "455--460",
    volume = 8,
    year = 1957,
    )
@ARTICLE(eM57a,
    author = "Ernest T. Moynahan",
    title = "On a problem of {M}.~{H}. {S}tone",
    journal = "Acta Math. Acad. Sci. Hungar.",
    pages = "455--460",
    volume = 8,
    number = 2,
    year = 1957,
    note = "Russian translation available",
    language = "English",
    )
```

Tip   Actually, for complete portability of the database, you should type titles with each important word capitalized:

```
title = "On a Problem of {M}.~{H}. {S}tone",
```

The bibliographic style file used in this book, `amsplain.bst`, converts "Problem" to "problem", so it makes no difference; but some bibliographic style files may not. So to be on the safe side, you should capitalize all words that may have to be capitalized. I'll not do that in this book, however.

Tip   The author field may contain the names of two or more joint authors, for instance,

```
author = "Al Blue and John Brown and Thomas Gray",
```

The bibliographic style file will typeset multiple author names correctly in the bibliography.

10.1.3 Books

Entry type: BOOK
Required fields: author (or editor), title, publisher, year
Optional fields: edition, volume, series, number, address, month,
                         language, note

Examples:

11. George A. Menuhin, *Universal algebra*, D. van Nostrand, Princeton-Toronto-London-Melbourne, 1968.
12. George A. Menuhin, *Universal algebra*, second ed., University Series in Higher Mathematics, vol. 58, D. van Nostrand, Princeton-Toronto-London-Melbourne, March 1968 (English), no Russian translation.

typed as

```
@BOOK(gM68,
    author = "George A. Menuhin",
    title = "Universal Algebra",
    publisher = "D.~van Nostrand",
    address = "Princeton-Toronto-London-Melbourne",
    year = 1968,
    )
```

```
@BOOK(gM68a,
    author = "George A. Menuhin",
    title = "Universal Algebra",
    publisher = "D.~van Nostrand",
    address = "Princeton-Toronto-London-Melbourne",
    year = 1968,
    series = "University Series in Higher Mathematics",
    volume = 58,
    edition = "Second",
    month = mar,
    note = "no Russian translation",
    language = "English",
    )
```

Second variant, with editor:

15. Robert S. Prescott (ed.), *Universal algebra*, D. van Nostrand, Princeton-Toronto-London-Melbourne, 1968.

typed as

```
@BOOK(rP68,
    editor = "Robert S. Prescott",
    title = "Universal Algebra",
    publisher = "D.~van Nostrand",
    address = "Princeton-Toronto-London-Melbourne",
    year = 1968,
    )
```

10.1.4   *Conference proceedings and collections*

Entry type: INPROCEEDINGS
Required fields: author, title, booktitle, year
Optional fields: address, editor, volume, series, number, organization, publisher, month, note, pages, language

Examples:

7. Peter A. Konig, *Composition of functions*, Proceedings of the Conference on Universal Algebra, 1970.
8. Peter A. Konig, *Composition of functions*, Proceedings of the Conference on Universal Algebra (Kingston, ON) (G. H. Birnbaum, ed.), vol. 7, Cana-

dian Mathematical Society, Queen's Univ., December 1970, available from
the Montreal office, pp. 1–106 (English).

typed as

```
@INPROCEEDINGS(pK69,
    author = "Peter A. Konig",
    title = "Composition of functions",
    booktitle = "Proceedings of the Conference on
     Universal Algebra",
    year = 1970,
    )
@INPROCEEDINGS(pK69a,
    author = "Peter A. Konig",
    title = "Composition of functions",
    booktitle = "Proceedings of the Conference on
     Universal Algebra",
    address = "Kingston, ON",
    publisher = "Queen's Univ.",
    organization = "Canadian Mathematical Society",
    editor = "G. H. Birnbaum",
    pages = "1--106",
    volume = 7,
    year = 1970,
    month = dec,
    note = "available from the Montreal office",
    language = "English",
    )
```

The address provides the location of the meeting. The address of the publisher,
if needed, should be typed in the publisher field, and the address of the organization, if needed, should be typed in the organization field.

Entry type: INCOLLECTION
Required fields: author, title, booktitle, publisher, year
Optional fields: editor, volume, series, number, address, edition, month,
                     note, pages, language

Examples:

1. Henry H. Albert, *Free torsoids*, Current Trends in Lattices, D. van Nostrand, 1970.
2. Henry H. Albert, *Free torsoids*, Current Trends in Lattices (George Burns,

ed.), vol. 2, D. van Nostrand, Princeton-Toronto-London-Melbourne, January 1970, new edition is due next year, pp. 173–215 (German).

typed as

```
@INCOLLECTION(hA70,
    author = "Henry H. Albert",
    title = "Free torsoids",
    booktitle = "Current Trends in Lattices",
    publisher = "D.~van Nostrand",
    year = 1970,
    )
@INCOLLECTION(hA70a,
    author = "Henry H. Albert",
    editor = "George Burns",
    title = "Free torsoids",
    booktitle = "Current Trends in Lattices",
    publisher = "D.~van Nostrand",
    address = "Princeton-Toronto-London-Melbourne",
    pages = "173--215",
    volume = 2,
    year = 1970,
    month = jan,
    note = "new edition is due next year",
    language = "German",
    )
```

The address field contains the address of the publisher.

10.1.5 Theses

Entry type: MASTERSTHESIS or PHDTHESIS
Required fields: author, title, school, year
Optional fields: type, address, month, note, pages (only for Ph.D. thesis)

Examples:

3. Soo-Key Foo, *Lattice Constructions*, Ph.D. thesis, University of Winnebago, 1990.

4. Soo-Key Foo, *Lattice Constructions*, Ph.D. dissertation, University of Winnebago, Winnebago, MN, December 1990, final revision not yet available.

typed as

```
@PHDTHESIS(sF90,
    author = "Soo-Key Foo",
    title = "Lattice Constructions",
    school = "University of Winnebago",
    year = 1990,
    )
@PHDTHESIS(sF90a,
    author = "Soo-Key Foo",
    title = "Lattice Constructions",
    school = "University of Winnebago",
    address = "Winnebago, MN",
    year = 1990,
    month = dec,
    note = "final revision not yet available",
    type = "Ph.D. dissertation",
    )
```

The type field if used takes the place of the phrase "Ph.D. thesis" (or "Master's thesis").

10.1.6 *Technical reports*

Entry type: TECHREPORT
Required fields: author, title, institution, year
Optional fields: type, number, address, month, note

Examples:

5. Grant H. Foster, *Computational complexity in lattice theory*, tech. report, Carnegie Mellon University, 1986.
6. Grant H. Foster, *Computational complexity in lattice theory*, Research Note 128A, Carnegie Mellon University, Pittsburgh, PA, December 1986, in preparation.

typed as

```
@TECHREPORT(gF86,
    author = "Grant H. Foster",
    title = "Computational complexity in lattice theory",
    institution = "Carnegie Mellon University",
    year = 1986,
    )
@TECHREPORT(gF86a,
```

```
    author = "Grant H. Foster",
    title = "Computational complexity in lattice theory",
    institution = "Carnegie Mellon University",
    year = 1986,
    type = "Research Note",
    address = "Pittsburgh, PA",
    number = "128A",
    month = dec,
    note = "in preparation",
    )
```

10.1.7 Manuscripts

Entry type: UNPUBLISHED
Required fields: author, title, note
Optional fields: month, year

Examples:

9. William A. Landau, *Representations of complete lattices*, manuscript, 55 pages.
10. William A. Landau, *Representations of complete lattices*, manuscript, 55 pages, December 1975.

typed as

```
@UNPUBLISHED(wL75,
    author = "William A. Landau",
    title = "Representations of complete lattices",
    note = "manuscript, 55 pages",
    )
@UNPUBLISHED(wL75a,
    author = "William A. Landau",
    title = "Representations of complete lattices",
    year = 1975,
    note = "manuscript, 55 pages",
    month = dec,
    )
```

10.1.8 Other entry types

Other entry types are listed below:

Entry type: BOOKLET
Required field: title
Optional fields: author, howpublished, address, month, year, note

Entry type: INBOOK
Required fields: author or editor, title, chapter and/or pages, publisher
               year
Optional fields: volume, series, number, type, address, edition, month,
               pages, language, note

Entry type: MANUAL
Required field: title
Optional fields: author, organization, address, edition, month, year, note

Entry type: MISC
Required field: at least one of the optional fields below must be present
Optional fields: author, title, howpublished, month, year, note, pages

Entry type: PROCEEDINGS
Required fields: title, year
Optional field: editor, volume, series, number, address, organization,
               publisher, month, note

10.1.9 *Abbreviations*

You may have noticed the field month = dec in the examples. This is an example
of an abbreviation. Most BIBTEX style files, including the $\mathcal{AMS}$ style files, come
with abbreviations for the months: jan, feb, ... , dec. When an abbreviation is
used, it's not in quotes (").

The name of the abbreviation, such as dec, is a string of characters that starts
with a letter, does not contain spaces, =, comma, or any of the special characters
of section 2.4.4.

An abbreviation is defined with the command @STRING. Example:

```
@STRING{au = "Algebra Universalis"}
```

A string definition can be placed anywhere in a bib file, as long as it precedes the
string's first use.

The $\mathcal{AMS}$ supplies the mrabbrev.bib file containing abbreviations for many
mathematical journals. This is an important file since it contains standard abbreviated forms. Based on this file, you can make your own myabbrev.bib file[1] containing all the journals you reference with whatever labels you find easy to remember.

If you use this scheme, the command specifying the bib files will always look
like

---
[1]This is a good idea, especially since the original may be too large for some systems to handle.

```
\bibliography{myabbrev,... }
```

(See section 10.2.1.)

10.2   *Using BIBTEX*

In section 10.1, you learned how to create bibliographic database files (the sample
bib files are `template.bib` and `sampart1.bib` in the `ftp` directory—see page 4).
In this section, you learn how to use BIBTEX to process these files.

10.2.1   *The sample files*

Create `sampart1.tex` from the `sampart.tex` sample article by replacing the bibliography with the two lines

```
\bibliographystyle{amsplain}
\bibliography{sampart1}
```

You can also find `sampart1.tex` in the `ftp` directory—see page 4. The first line
names the bst file and the second line names the bibliographic database files used;
in this case there is only one, `sampart1.bib`.

The `amsplain.bst` file is the $\mathcal{AMS}$ plain bibliographic style file; it comes with
the $\mathcal{AMS}$ distribution (see Appendix G). Use `amsplain.bst` if you want the bibliographic items be formatted as presented in this book. The LaTeX distribution
also comes with a number of other bibliographic style files.

The `sampart1.bib` file is the bibliographic database for `sampart1.tex`. It
is the file referred to in the `\bibliography` command above. Its contents are as
follows:

```
@BOOK(gM68,
    author = "George A. Menuhin",
    title = "Universal Algebra",
    publisher = "D.~van Nostrand",
    address = "Princeton-Toronto-London-Melbourne",
    year = 1968,


@BOOK(fR82,
    author = "Ferenc R. Richardson",
    title = "General Lattice Theory",
    edition = "Expanded and Revised",
    language = "Russian",
    publisher = "Mir",
    address = "Moscow",
```

```
        year = 1982,
   )

   @ARTICLE(eM57,
        author = "Ernest T. Moynahan",
        title = "On a problem of {M}.~{H}. {S}tone",
        journal = "Acta Math. Acad. Sci. Hungar.",
        pages = "455--460",
        volume = 8,
        year = 1957,
   )

   @ARTICLE(eM57a,
        author = "Ernest T. Moynahan",
        title = "Ideals and congruence relations in lattices.~{II}",
        journal = "Magyar Tud. Akad. Mat. Fiz. Oszt. K{\"{o}}zl.",
        language = "Hungarian",
        pages = "417--434",
        volume = 7,
        year = 1957,
   )

   @PHDTHESIS(sF90,
        author = "Soo-Key Foo",
        title = "Lattice Constructions",
        school = "University of Winnebago",
        address = "Winnebago, MN",
        month = dec,
        year = 1990,
   )
```

Type in sampart1.bib or copy it from the ftp directory.

10.2.2   The setup

Before you start BIBT_EX, make sure that everything is set up properly, as described in this section.

You specify which references go into the bibliography with the \cite commands; they work as discussed in section 6.5.1. If you choose to include a reference that is not mentioned in the text, pull it in with a \nocite command. Example:

\cite{pK57}

includes and cites pK57, while

\nocite{pK57}

includes and but does not cite pK57. In either case, one of the bib files specified in the argument of the \bibliography command must contain an entry with the label pK57. The

\nocite{*}

command includes *all* the items from the bibliographic databases specified with the \bibliography command.

The document must contain instructions naming the bib files to be used and specifying the style for the bibliography. For instance, the sampart1.tex sample article contains the lines:

\bibliographystyle{amsplain}
\bibliography{sampart1}

The \bibliographystyle command names amsplain.bst as the bst file and the \bibliography command names the bib file sampart1.bib. Alternatively, one could write

\bibliography{myabbrev,gg,lattice,sampart1}

where myabbrev.bib contains user-defined abbreviations, gg.bib contains personal articles, lattice.bib contains lattice theoretical articles by other authors, and sampart1.bib contains additional references needed for sampart1.

It is important to make sure that the bst file, the bib file(s), and the LaTeX document(s) are in subdirectories where BIBT_EX can find them; you might want myabbrev.bib and lattice.bib in one "central" location, while sampart1.bib would naturally be placed in the same directory as the document itself.

If you are just starting out, simply copy all of them into one subdirectory; later, you may want to look for a more permanent solution.

10.2.3   The four steps of BIBT_EXing

The following steps produce a bibliography; use the sampart1.tex sample article as an example:

Step 1   Check that BibTeX, the LaTeX document, and the bib files are placed in the appropriate directories (see the comment at the end of section 10.2.2).

Step 2   Typeset the document sampart1.tex to get a fresh aux file; this step is illustrated in Figure 10.1.

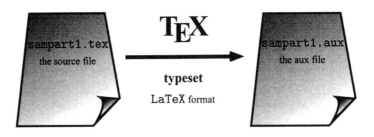

Figure 10.1: Using BibTeX, Step 2

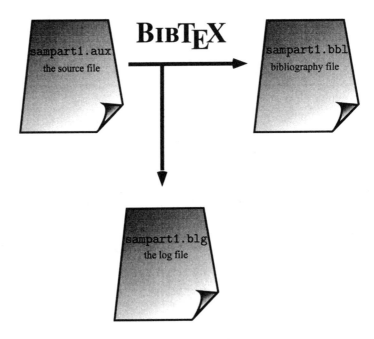

Figure 10.2: Using BibTeX, Step 3

Step 3 Run BibTeX on the `sampart1.aux` file by invoking it with the argument `sampart1`. If BibTeX can't find a crucial file—for example, the `bst` file—it'll stop. The reason why it stopped will be shown on the monitor and written in a `blg` (bibliographic log) file, `sampart1.blg`. Correct the errors and run BibTeX again. A successful run creates a `bbl` (bibliography) file, `sampart1.bbl`, in addition to `sampart1.blg`. This step is illustrated in Figure 10.2.

Step 4 Typeset the LaTeX document `sampart1.tex` **twice**.

10.2.4 *The files of BibTeX*

To illustrate the process, complete the following steps using `sampart1.tex`:

Step 1 Start fresh by deleting the `aux`, `blg`, and `bbl` files, if they are present.

Step 2 Typeset the article `sampart1.tex` to get an aux file (see Figure 10.1). (Notice that the log file contains warnings about missing references and a number of lines not relevant to the current discussion.) The lines in the aux file containing the bibliographic information are:

```
\citation{fR82}
\citation{gM68}
\citation{eM57}
\citation{sF90}
\citation{eM57a}
\bibstyle{amsplain}
\bibdata{sampart1}
```

Each `\citation` in this file corresponds to a `\cite` (or `\nocite`) in the article. The lines

```
\bibliographystyle{amsplain}
\bibliography{sampart1}
```

of the article `sampart1.tex` are translated into

```
\bibstyle{amsplain}
\bibdata{sampart1}
```

in the aux file.

Step 3 Now run BibTeX on the `sampart1.aux` file by invoking it with the argument `sampart1` (see Figure 10.2). BibTeX generates two new files, `sampart1.blg` and `sampart1.bbl`. Look at `sampart1.blg`:

```
This is BibTeX, C Version 0.99c
The top-level auxiliary file: article1.aux
The style file: amsplain.bst
Database file #1: sampart1.bib
```

At present, this blg file does not contain much important information; however, if there are errors, they'll be listed in this file.

The sampart1.bbl file, in which BIBTEX created the thebibliography environment as described in section 6.5.1, is more interesting:

```
\providecommand{\bysame}{\leavevmode\hbox to 3em%
{\hrulefill}\thinspace}
\begin{thebibliography}{1}

\bibitem{sF90}
Soo-Key Foo, \emph{Lattice Constructions}, Ph.D. thesis,
University of Winnebago, Winnebago, MN, December 1990.

\bibitem{gM68}
George~A. Menuhin, \emph{Universal algebra}, D.~van Nostrand,
Princeton-Toronto-London-Melbourne, 1968.

\bibitem{eM57a}
Ernest~T. Moynahan, \emph{Ideals and congruence relations in
lattices.~{II}}, Magyar Tud. Akad. Mat. Fiz. Oszt. K{\"{o}}zl.
\textbf{7} (1957), 417--434 (Hungarian).

\bibitem{eM57}
\bysame, \emph{On a problem of {M}.~{H}. {S}tone, Acta Math.
Acad. Sci. Hungar. \textbf{8} (1957), 455--460.

\bibitem{fR82}
Ferenc~R. Richardson, \emph{General lattice theory}, expanded
and revised ed., Mir, Moscow, 1982 (Russian).

\end{thebibliography}
```

Observe how the nonbreakable spaces are provided for the authors.

Step 4   Now typeset the document sampart1.tex. In the typeset form, the "References" section appears (it was constructed from the bbl file); however, in the new log file, you still get the warnings about missing bibliographic references. The new aux file contains five lines of interesting new information:

```
\bibcite{sF90}{1}
\bibcite{gM68}{2}
\bibcite{eM57a}{3}
\bibcite{eM57}{4}
\bibcite{fR82}{5}
```

which identifies the bibliographic cross-reference symbol sF90 (see line 1 above—
the symbol designates Foo's thesis in sampart1.bib) with the number 1, and so
on. Now typeset sampart1.tex again, and all the references will be correctly given
in the typeset article.

Observe:

1. The crucial step in Step 3 (see Figure 10.2), the running of the BIBTEX program, gives you different error messages and obeys different rules (compared
with LATEX—see section 10.2.5).
2. The sampart1.bbl file was created by BIBTEX. It'll not be changed by running
LATEX again (but it may be edited).

10.2.5 BIBTEX *rules and messages*

Rule ■ BIBTEX and %
You cannot comment out a field with % in BIBTEX version 0.99.

For example, the following item

```
@ARTICLE(eM57,
    author = "Ernest T. Moynahan",
    title = "On a problem of {M}.~{H}. {S}tone",
    journal = "Acta Math. Acad. Sci. Hungar.",
    % pages = "455--460",
    volume = 8,
    year = 1957,
    )
```

will cause BIBTEX to give the error message:

```
You're missing a field name---line 23 of file sampart1.bib
  :
  :    % pages = "455--460",
(Error may have been on previous line)
I'm skipping whatever remains of this entry
Warning--missing year in eM57
Warning--missing pages in eM57
(There was 1 error message)
```

Recall that BIBTEX ignores field names it can't recognize. So changing the
field name pages, say, to pages-comment will not give an error message. However,
this "comments out" a required field, so you get a warning:

```
Warning--missing pages in eM57
```

Rule ■ BIBT_EX field names
Do not abbreviate field names.

For instance, if you abbreviate volume to vol as in:

```
@ARTICLE(eM57,
    author = "Ernest T. Moynahan",
    title = "On a problem of {M}.~{H}. {S}tone",
    journal = "Acta Math. Acad. Sci. Hungar.",
    pages = "455--460",
    vol = 8,
    year = 1957,
    )
```

then the volume is simply ignored. This item will print:

12. Ernest T. Moynahan, *On a problem of M. H. Stone*, Acta Math. Acad. Sci. Hungar. (1957), 455–460.

Rule ■ BIBT_EX field termination
Make sure that every line of an entry, except possibly the last, is terminated with a comma.

If not, you get an error message:

```
I was expecting a ',' or a ')'---line 6 of file sampart1.bib
 :
 :     address = "Princeton-Toronto-London-Melbourne",
(Error may have been on previous line)
I'm skipping whatever remains of this entry
Warning--missing year in gM68
```

Rule ■ BIBT_EX field value termination
Make sure that the field value is properly terminated.

You should be careful not to drop a double quote. If you drop the closing " in line 11 of the bib file, you get the error message:

```
I was expecting a ',' or a ')'---line 12 of file sampart1.bib
 :     edition = "
 :                    Expanded and Revised",
I'm skipping whatever remains of this entry
Warning--missing publisher in fR82
Warning--missing year in fR82
```

If instead you drop the opening double quote in the same line, you get the error message:

```
Warning--string name "general" is undefined
--line 11 of file sampart1.bib
I was expecting a ',' or a ')'---line 11 of file sampart1.bib
 :     title =  general
 :                    Lattice Theory",
I'm skipping whatever remains of this entry
Warning--missing title in fR82
Warning--missing publisher in fR82
Warning--missing year in fR82
(There was 1 error message)
```

BIBTEX assumed that general is a string, since it was not preceded by ".

The obvious conclusion is that one has to be very careful about typing BIBTEX entries. There may be special tools available for your computer system that assist in the collection of bibliographic data (see Chapter 12 in [45]). If you do not have access to such tools, I recommend that you use the template.bib file that contains templates of often-used bibliographic entry types. Copy the forms you need and fill in the blanks; this will avoid typos that lead to error messages.

10.2.6   Concluding comments

There is a lot more to BIBTEX than what was covered in this chapter. For example, BIBTEX's algorithm to alphabetize names is fairly complicated. Some names are complex: where does "von Neumann" go (with the "v"s or the "N"s), or how about "Miguel Lopez Fernandez" go (with the "L"s if the family name is "Lopez Fernandez")?

The document [38] has many helpful hints, including a clever hack to order bibliographic items correctly even when BIBTEX does not have enough information on how to do this. Chapter 13 of *The LaTeX Companion* has a long discussion of BIBTEX. It also contains a long list of bibliographic style files.

Some time after the publication of this book, a new version of BIBTEX is scheduled for release (see [39]). Mostly it'll add new features, however, there may be changes that will affect the names of some internal commands, such as \bibstyle.

MakeIndex

To assist the LaTeX user in making a long index, Pehong Chen wrote the program *MakeIndex* (see [9]). For short indexes, you can easily do without it (see section 6.5.2).

11.1 Preparing the document

It is easy to create an index with the help of *MakeIndex*: just mark the index entries in the source file with \index commands, and then let LaTeX and *MakeIndex* do the rest.

- In the preamble, include the lines:

```
\usepackage{makeidx}
\makeindex
```

  (If you use an $\mathcal{AMS}$ document class, *do not include the first line*; if you do, you get an error message.)
- Where you want the index printed in the document, normally in the back matter, type the line

```
\printindex
```

- Mark all index entries with the \index command.

This procedure will be illustrated with the intrarti.tex article (in your ftp directory—see page 4), which modifies intrart.tex by inserting a number of index entries.

The picture below shows the index entries created by Entries 1–8.

Entry 1 Rewrite the line:

```
\begin{theorem}
```

as follows:

```
\begin{theorem}\index{Main Theorem}
```

Entries 2 and 3 Type the commands

```
\index{pistar@$\Pi^{*}$ construction}%
\index{Main Theorem!exposition|(}%
```

so that they follow the line

```
\section{The $\Pi^{*}$ construction} \label{S:P*}
```

Entry 4 Type

```
\index{<@$\langle \dots, 0, \dots, \overset{i}{d}, \dots, 0,
      \dots \rangle$|textbf}
```

to follow the line

```
\begin{notation}
```

Entry 5 Rewrite the line

```
See also Ernest~T.
Moynahan~\cite{eM57a}.
```

as follows:

```
See also Ernest~T.
\index{Moynahan, Ernest~T.}%
Moynahan~\cite{eM57a}.
```

Entries 6 to 8 The following three index items

```
\index{lattice}%
\index{lattice!distributive}%
\index{lattice!distributive!complete}%
```

should precede the line

```
\begin{theorem} \label{T:P*}
```

Index

$$4 \bullet\!\!\!\overline{\langle \dots, 0, \dots, \overset{i}{d}, \dots, 0, \dots \rangle}, \mathbf{1}, 2$$

Foo, Soo-Key, 2

$$6 \bullet\!\!\!\overline{\text{lattice}, 2}$$
$$7 \bullet\!\!\!\overline{\text{distributive}, 2}$$
$$8 \bullet\!\!\!\overline{\text{complete}, 2}$$

$$1 \bullet\!\!\!\overline{\text{Main Theorem}, 1}$$
$$3 \bullet\!\!\!\overline{\text{exposition}, 1\text{--}2}$$

Menuhin, George A., 3
Moynahan, Ernest T., 2, 3

$$5 \bullet\!\!\!\overline{}$$
$$2 \bullet\!\!\!\overline{\Pi^{*} \text{ construction}, 1}$$

The picture below shows the index entries created by Entries 9–14.

Entry 9   Insert the item

```
\index{<@$\langle \dots, 0, \dots, \overset{i}{d}, \dots, 0,
    \dots \rangle$}%
```

to follow the line

```
we get:
```

Entry 10   Type

```
\index{Main Theorem!exposition|)}
```

to follow the line

```
hence $\Theta = \iota$.
```

Entry 11   Insert

```
\index{Foo, Soo-Key}%
```

to follow the line

```
\bibitem{sF90}
```

Entry 12   Type

```
\index{Menuhin, George~A.}%
```

to follow the line

```
\bibitem{gM68}
```

Entries 13 and 14   Insert the following item twice

```
\index{Moynahan, Ernest~T.}%
```

first, to follow the line

```
\bibitem{eM57}
```

and then to follow the line

```
\bibitem{eM57a}
```

Index

$$9 \quad \underline{\langle \dots, 0, \dots, \overset{i}{d}, \dots, 0, \dots \rangle}, \mathbf{1}, \mathbf{2}$$

$$11 \quad \underline{\text{Foo, Soo-Key}, 2}$$

lattice, 2

distributive, 2

complete, 2

Main Theorem, 1

$$10 \quad \underline{\text{exposition}, 1\text{–}2}$$
$$12 \quad \underline{\text{Menuhin, George A., 3}}$$
$$13, 14 \quad \underline{\text{Moynahan, Ernest T., 2, 3}}$$

Π^* construction, 1

These \index commands produce the index for the intrarti.tex article shown in Figure 11.1. Observe that there are only 13 entries in the index, whereas you typed 14 index entries into intrarti.tex. The reason is that the last two entries for Moynahan (Entries 13 and 14) happen to fall on page 2, so only one page number shows up in the index.

Index

$$\langle \ldots, 0, \ldots, \overset{i}{d}, \ldots, 0, \ldots \rangle, \mathbf{1}, 2$$

Foo, Soo-Key, 2

lattice, 2
> distributive, 2
>> complete, 2

Main Theorem, 1
> exposition, 1–2
Menuhin, George A., 3
Moynahan, Ernest T., 2, 3

Π^* construction, 1

Figure 11.1: A sample index

11.2 *Index entries*

There are a few major types of index entries. They are discussed in this section, illustrated by the entries in section 11.1.

Simple entries

The entry in the index:

Foo, Soo-Key, 2

was created as Entry 11 with the command:

`\index{Foo, Soo-Key}`

This is the simplest form of an index entry:

`\index{`*entry_text*`}`

Subentries

The item

lattice, 2

was created as Entry 6 with the command:

\index{lattice}

It has a subentry:

lattice, 2
    distributive, 2

which was created as Entry 7 with the command:

\index{lattice!distributive}

In fact, there is also a subsubentry:

lattice, 2
    distributive, 2
        complete, 2

which was created as Entry 8 with the command:

\index{lattice!distributive!complete}

In general, you mark a subentry with

\index{ *entry* ! *subentry* }

and a subsubentry with

\index{ *entry* ! *subentry* ! *subsubentry* }

Modifiers

The entry

\index{Moynahan, Ernest T.|textbf}

produces the bold **2** page number for the Moynahan entry. The command whose name follows | (the command name textbf) is applied to the page number. Observe that the modifier becomes a command, and the argument of the command is the page number. For instance, if you want a Large, bold page number, define the command:

\newcommand{\LB}[1]{\textbf{\Large #1}}

and modify the entry thus:

```
\index{entry|LB}
```

You can also modify entries to make *page ranges*.

Main Theorem, 1
    exposition, 1–2

The latter index entry has a page range; it was created with Entries 3 and 10:

```
\index{Main Theorem!exposition|(}
\index{Main Theorem!exposition|)}
```

Separate the (sub)entry from its modifier with |, open the page range with (, and close it with).

Modifiers can also be combined. The index entries

```
\index{Main Theorem!exposition|(textbf}
\index{Main Theorem!exposition|)textbf}
```

produce a bold page range.

Placement control

Observe the entry that occurs twice (once with a `textbf` modifier):

```
\index{<@$\langle \dots, 0, \dots, \overset{i}{d}, \dots, 0,
        \dots \rangle$}
```

This pair of entries produce the index entry

$$\langle \dots, 0, \dots, \overset{i}{d}, \dots, 0, \dots \rangle, \mathbf{1}, 2$$

To place this entry at the correct place in the index, use a *placement specifier*. The general form is

```
\index{placement@entry}
```

When the index items are sorted, the value of *placement* is used in the sorting. Two typical uses are as follows:

```
\index{itshape@\com{itshape}}
```

(where \com{*command_name*} prints *command_name*; see section 6.5.2) places the \itshape command in the index, and

```
\index{Doob, Michael@Michael Doob}
```

for names. (This is how I place names in the index of this book. This is quite unconventional, since by tradition, "Michael Doob" should be entered as "Doob, Michael". This convention developed since most indexing programs lack the placement control capability of *MakeIndex.*)

Special characters

Since the characters !, @, and | have special meanings, you must *quote* them if they appear in an index entry: "!, "@, and "|. Since this makes " special also, it must be quoted as well: "". The only exception is \" (to allow for the accented character \"{a}); in this context, " does not quote the next character.

Example: to produce the index entry $|A|$, enter it as follows:

```
\index{"|A"|@$"|A"|$}
```

Cross-references

It is easy to make a reference to another index entry; for instance, to list "distributive lattice", yet refer to "lattice, distributive", the command is

```
\index{distributive lattice|see{lattice, distributive}}
```

This command can be placed anywhere in the document. This produces the entry:

distributive lattice, *see* lattice, distributive

Tip   Put all index cross-references in one place in the document, so they are easy to keep track of.

Where to place the entry?

The principle is simple:

Rule   ■ Placement of the index entry

- The index entry should reference the correct page.
- The index entry should not introduce unwanted space.

For example, avoid the following:

```
Let $L$ be a distributive lattice
\index{lattice}
\index{distributive lattice}
that is strongly complete.
```

This may put an unwanted space following the word "lattice".

Note the placement of the index entries in the examples; in each case I placed them as close to the referenced item as I could. If you place an index entry on a separate line, use % to comment out unwanted spaces (including the end-of-line character), as in

```
Let $L$ be a distributive lattice
\index{lattice}%
\index{distributive lattice}%
that is strongly complete.
```

11.3 Processing the index entries

Now you are ready to create the index:

Step 1   Typeset the document intrarti.tex (see Figure 11.2).

Step 2   Run the *MakeIndex* program on intrarti.idx (you are asked to name the document without an extension) (see Figure 11.3).

Step 3   Typeset the document intrarti.tex again.
On page 3 of the typeset document, you'll find the index.

Let's look at this process in detail. In Step 1 (see Figure 11.2), LaTeX creates the intrarti.idx file:

```
\indexentry{Main Theorem}{1}
\indexentry{pistar@$\Pi^{*}$ construction}{1}
\indexentry{Main Theorem!exposition|(}{1}
\indexentry{<@$\langle \dots, 0, \dots, \overset{i}{d}, \dots, 0,
    \dots \rangle$|textbf}{1}
\indexentry{Moynahan, Ernest~T.}{2}
\indexentry{lattice}{2}
\indexentry{lattice!distributive}{2}
\indexentry{lattice!distributive!complete}{2}
\indexentry{<@$\langle \dots, 0, \dots, \overset{i}{d}, \dots, 0,
    \dots \rangle$}{2}
\indexentry{Main Theorem!exposition|)}{2}
\indexentry{Foo, Soo-Key}{2}
\indexentry{Menuhin, George~A.}{2}
\indexentry{Moynahan, Ernest~T.}{2}
\indexentry{Moynahan, Ernest~T.}{2}
```

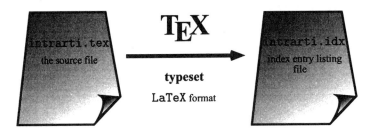

Figure 11.2: Using *MakeIndex*, Step 1

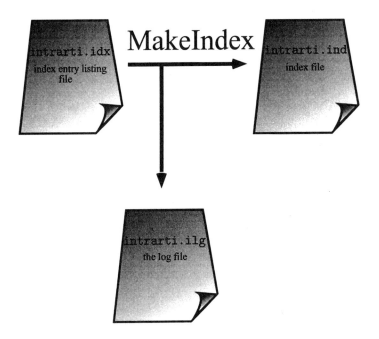

Figure 11.3: Using *MakeIndex*, Step 2

In Step 2 (see Figure 11.3), *MakeIndex* processes `intrarti.idx` (*MakeIndex* is invoked with the argument `intrarti`) and creates the index file `intrarti.ind`:

```
\begin{theindex}

  \item $\langle \dots, 0, \dots, \overset{i}{d}, \dots, 0,
     \dots \rangle$, \textbf{1}, 2
```

```
\indexspace

\item Foo, Soo-Key, 2

\indexspace

\item lattice, 2
  \subitem distributive, 2
    \subsubitem complete, 2

\indexspace

\item Main Theorem, 1
  \subitem exposition, 1--2
\item Menuhin, George~A., 2
\item Moynahan, Ernest~T., 2

\indexspace

\item $ \Pi^{*} $ construction, 1
```

`\end{theindex}`

which contains a theindex environment with all the index entries. The

`\printindex`

command acts as an \include, read intrarti.ind during the next typesetting.
    MakeIndex also produces the index log file intrarti.ilg:

```
This is MakeIndex, portable version 2.12 [26-May-1993].
Scanning input file intrarti.idx....done
(14 entries accepted, 0 rejected).
Sorting entries....done (53 comparisons).
Generating output file intrarti.ind....done
(27 lines written, 0 warnings).
Output written in intrarti.ind.
Transcript written in intrarti.ilg.
```

It is important to understand that in Step 1, LaTeX does not process the index
entries; it writes out the arguments of the \index commands in the source file
as arguments of \indexentry commands "verbatim" (that is, with no change) to
the idx file. *MakeIndex* then processes the idx file by removing the double quote
marks for the special characters, sorting the entries by position, and collating the

numbers. The resulting ind file is a normal LaTeX source file (which may be edited) that is included in the original document the next time you run LaTeX.

11.4   *Rules*

There are some simple rules to keep in mind when entering index items.

Rule ■ Spaces in \index
Do not leave unnecessary spaces in the argument of an \index command.

\index{item}, \index{⊔item}, \index{item⊔}

are three different entries.

There are options that instruct *MakeIndex* to ignore such spaces; however, it is better to do it correctly in the first place.

Rule ■ Spacing rules of LaTeX
The spacing rules of LaTeX (section 2.2.1) do not apply; *MakeIndex* does not follow these rules when it does the sorting of the index items.

Rule ■ Positioning
In \index{*position*@*item*}, the *position* is both space and case sensitive.

For instance,

\index{alpha@α}
\index{Alpha@α}
\index{ALPHA@α}

represent three different items.

Rule ■ Positioning and subitems
If you use a *position* clause, it becomes part of the item, and must be used for subitems.

So if the item is designated by \index{Doob@Michael Doob}, then a subitem must have the form

```
\index{Doob@Michael Doob!the life and times of}
```

and **not**

```
\index{Michael Doob!the life and times of}
```

Rule ■ amsmath

If you use the amsmath package, do not place an \index command in an equation
or in any $\mathcal{AMS}$ multiline math environment.

If you violate this rule, you may get an error message:

```
! TeX capacity exceeded, sorry [input stack size=200].
\restorecounters@ ...ecounters@
                                  \@empty
l.9     \end{equation}
```

```
If you really absolutely need more capacity,
you can ask a wizard to enlarge me.
```

Rule ■ Braces

In every entry, the braces must be balanced.

For instance, an entry that places \{ at the position leftbrace cannot be typed as

```
\index{leftbrace@\{}
```

LaTeX will give the error message:

```
Runaway argument?
{leftbrace@\{}
! Paragraph ended before \@wrindex was complete.
```

There are many ways to correct this entry. Perhaps the simplest is to define

```
\newcommand{\printleftbrace}{\{}
```

and rewrite the entry:

```
\index{leftbrace@\printleftbrace}
```

This produces the index entry:

$$\{, 1$$

positioned alphabetically as if it were "leftbrace".

There is, of course, a lot more to *MakeIndex* than what was given in this short introduction, but what we have covered here will do for most documents. See Pehong Chen and Michael A. Harrison [9] for more detail. Chapter 12 of *The LaTeX Companion* covers *MakeIndex* in great detail, including multiple indexes and customization.

11.5   *Glossary*

Producing a glossary is very similar to an index.

Instead of the \index and \makeindex commands, use the \glossary and \makeglossary commands, respectively. The glossary entries are written in the glo file.

Math symbol tables

A.1 Hebrew letters

Type:	Print:	Type:	Print:
\aleph	א	\beth	ב
\daleth	ד	\gimel	ג

All symbols but \aleph need the amssymb package.

A.2   *Greek characters*

Type:	Print:	Type:	Print:	Type:	Print:
\alpha	α	\beta	β	\gamma	γ
\digamma	$\digamma$	\delta	δ	\epsilon	ϵ
\varepsilon	ε	\zeta	ζ	\eta	η
\theta	θ	\vartheta	ϑ	\iota	ι
\kappa	κ	\varkappa	$\varkappa$	\lambda	λ
\mu	μ	\nu	ν	\xi	ξ
\pi	π	\varpi	ϖ	\rho	ρ
\varrho	ϱ	\sigma	σ	\varsigma	ς
\tau	τ	\upsilon	υ	\phi	ϕ
\varphi	φ	\chi	χ	\psi	ψ
\omega	ω				

\digamma and \varkappa require the amssymb package.

Type:	Print:	Type:	Print:
\Gamma	Γ	\varGamma	$\varGamma$
\Delta	Δ	\varDelta	$\varDelta$
\Theta	Θ	\varTheta	$\varTheta$
\Lambda	Λ	\varLambda	$\varLambda$
\Xi	Ξ	\varXi	$\varXi$
\Pi	Π	\varPi	$\varPi$
\Sigma	Σ	\varSigma	$\varSigma$
\Upsilon	Υ	\varUpsilon	$\varUpsilon$
\Phi	Φ	\varPhi	$\varPhi$
\Psi	Ψ	\varPsi	$\varPsi$
\Omega	Ω	\varOmega	$\varOmega$

All symbols whose name begins with var need the amsmath package.

A.3   LaTeX *binary relations*

Type:	Print:	Type:	Print:
\in	$\in$	\ni	$\ni$
\leq	$\leq$	\geq	$\geq$
\ll	$\ll$	\gg	$\gg$
\prec	$\prec$	\succ	$\succ$
\preceq	$\preceq$	\succeq	$\succeq$
\sim	$\sim$	\cong	$\cong$
\simeq	$\simeq$	\approx	$\approx$
\equiv	$\equiv$	\doteq	$\doteq$
\subset	$\subset$	\supset	$\supset$
\subseteq	$\subseteq$	\supseteq	$\supseteq$
\sqsubseteq	$\sqsubseteq$	\sqsupseteq	$\sqsupseteq$
\smile	$\smile$	\frown	$\frown$
\perp	$\perp$	\models	$\models$
\mid	$\mid$	\parallel	$\parallel$
\vdash	$\vdash$	\dashv	$\dashv$
\propto	$\propto$	\asymp	$\asymp$
\bowtie	$\bowtie$		
\sqsubset	$\sqsubset$	\sqsupset	$\sqsupset$
\Join	$\Join$		

The latter three symbols need the latexsym package.

A.4 $\mathcal{AMS}$ *binary relations*

Type:	Print:	Type:	Print:
\leqslant	⩽	\geqslant	⩾
\eqslantless	⪕	\eqslantgtr	⪖
\lesssim	≲	\gtrsim	≳
\lessapprox	⪅	\gtrapprox	⪆
\approxeq	≊		
\lessdot	⋖	\gtrdot	⋗
\lll	⋘	\ggg	⋙
\lessgtr	≶	\gtrless	≷
\lesseqgtr	⋚	\gtreqless	⋛
\lesseqqgtr	⪋	\gtreqqless	⪌
\doteqdot	≑	\eqcirc	≖
\circeq	≗	\fallingdotseq	≒
\risingdotseq	≓	\triangleq	≜
\backsim	∽	\thicksim	∼
\backsimeq	⋍	\thickapprox	≈
\preccurlyeq	≼	\succcurlyeq	≽
\curlyeqprec	⋞	\curlyeqsucc	⋟
\precsim	≾	\succsim	≿
\precapprox	⪷	\succapprox	⪸
\subseteqq	⊆	\supseteqq	⊇
\Subset	⋐	\Supset	⋑
\vartriangleleft	◁	\vartriangleright	▷
\trianglelefteq	⊴	\trianglerighteq	⊵
\vDash	⊨	\Vdash	⊩
\Vvdash	⊪		
\smallsmile	⌣	\smallfrown	⌢
\shortmid	∣	\shortparallel	∥
\bumpeq	≏	\Bumpeq	≎
\between	≬	\pitchfork	⋔
\varpropto	∝	\backepsilon	϶
\blacktriangleleft	◀	\blacktriangleright	▶
\therefore	∴	\because	∵

All symbols require the amssymb package.

A.5 $\mathcal{AMS}$ *negated binary relations*

Type:	Print:	Type:	Print:
\ne	$\ne$	\notin	$\notin$
\nless	$\nless$	\ngtr	$\ngtr$
\nleq	$\nleq$	\ngeq	$\ngeq$
\nleqslant	$\nleqslant$	\ngeqslant	$\ngeqslant$
\nleqq	$\nleqq$	\ngeqq	$\ngeqq$
\lneq	$\lneq$	\gneq	$\gneq$
\lneqq	$\lneqq$	\gneqq	$\gneqq$
\lvertneqq	$\lvertneqq$	\gvertneqq	$\gvertneqq$
\lnsim	$\lnsim$	\gnsim	$\gnsim$
\lnapprox	$\lnapprox$	\gnapprox	$\gnapprox$
\nprec	$\nprec$	\nsucc	$\nsucc$
\npreceq	$\npreceq$	\nsucceq	$\nsucceq$
\precneqq	$\precneqq$	\succneqq	$\succneqq$
\precnsim	$\precnsim$	\succnsim	$\succnsim$
\precnapprox	$\precnapprox$	\succnapprox	$\succnapprox$
\nsim	$\nsim$	\ncong	$\ncong$
\nshortmid	$\nshortmid$	\nshortparallel	$\nshortparallel$
\nmid	$\nmid$	\nparallel	$\nparallel$
\nvdash	$\nvdash$	\nvDash	$\nvDash$
\nVdash	$\nVdash$	\nVDash	$\nVDash$
\ntriangleleft	$\ntriangleleft$	\ntriangleright	$\ntriangleright$
\ntrianglelefteq	$\ntrianglelefteq$	\ntrianglerighteq	$\ntrianglerighteq$
\nsubseteq	$\nsubseteq$	\nsupseteq	$\nsupseteq$
\nsubseteqq	$\nsubseteqq$	\nsupseteqq	$\nsupseteqq$
\subsetneq	$\subsetneq$	\supsetneq	$\supsetneq$
\varsubsetneq	$\varsubsetneq$	\varsupsetneq	$\varsupsetneq$
\subsetneqq	$\subsetneqq$	\supsetneqq	$\supsetneqq$
\varsubsetneqq	$\varsubsetneqq$	\varsupsetneqq	$\varsupsetneqq$

All symbols but \ne require the amssymb package.

A.6 *Binary operations*

Type:	Print:	Type:	Print:
\pm	±	\mp	∓
\times	×	\cdot	·
\circ	∘	\bigcirc	◯
\div	÷	\diamond	◇
\ast	∗	\star	⋆
\cap	∩	\cup	∪
\sqcap	⊓	\sqcup	⊔
\wedge	∧	\vee	∨
\triangleleft	◁	\triangleright	▷
\bigtriangleup	△	\bigtriangledown	▽
\oplus	⊕	\ominus	⊖
\otimes	⊗	\oslash	⊘
\odot	⊙	\bullet	●
\dagger	†	\ddagger	‡
\setminus	\	\uplus	⊎
\wr	≀	\amalg	⨿
\lhd	◁	\rhd	▷
\unlhd	⊴	\unrhd	⊵
\dotplus	∔	\centerdot	·
\ltimes	⋉	\rtimes	⋊
\leftthreetimes	⋋	\rightthreetimes	⋌
\circleddash	⊝	\smallsetminus	∖
\barwedge	⊼	\doublebarwedge	⩞
\curlywedge	⋏	\curlyvee	⋎
\veebar	⊻	\intercal	⊺
\Cap	⋒	\Cup	⋓
\circledast	⊛	\circledcirc	⊚
\boxminus	⊟	\boxtimes	⊠
\boxdot	⊡	\boxplus	⊞
\divideontimes	⋇		
\And	&		

This table is divided into four parts. The first part contains the binary operations in LaTeX. The second part requires the latexsym package. The third part contains the $\mathcal{AMS}$ additions; they require the amssymb package. The symbol \And requires the amsmath package.

A.7   Arrows

Type:	Print:	Type:	Print:
\leftarrow	←	\rightarrow or \to	→
\longleftarrow	⟵	\longrightarrow	⟶
\Leftarrow	⇐	\Rightarrow	⇒
\Longleftarrow	⟸	\Longrightarrow	⟹
\leftrightarrow	↔	\longleftrightarrow	⟷
\Leftrightarrow	⇔	\Longleftrightarrow	⟺
\uparrow	↑	\downarrow	↓
\Uparrow	⇑	\Downarrow	⇓
\updownarrow	↕	\Updownarrow	⇕
\nearrow	↗	\searrow	↘
\swarrow	↙	\nwarrow	↖
\mapsto	↦	\longmapsto	⟼
\hookleftarrow	↩	\hookrightarrow	↪
\leftharpoonup	↼	\rightharpoonup	⇀
\leftharpoondown	↽	\rightharpoondown	⇁
\rightleftharpoons	⇌		
\leadsto	⇝		
\leftleftarrows	⇇	\rightrightarrows	⇉
\leftrightarrows	⇆	\rightleftarrows	⇄
\Lleftarrow	⇚	\Rrightarrow	⇛
\twoheadleftarrow	↞	\twoheadrightarrow	↠
\leftarrowtail	↢	\rightarrowtail	↣
\looparrowleft	↫	\looparrowright	↬
\upuparrows	⇈	\downdownarrows	⇊
\upharpoonleft	↿	\upharpoonright	↾
\downharpoonleft	⇃	\downharpoonright	⇂
\leftrightsquigarrow	↭	\rightsquigarrow	⇝
\multimap	⊸		
\nleftarrow	↚	\nrightarrow	↛
\nLeftarrow	⇍	\nRightarrow	⇏
\nleftrightarrow	↮	\nLeftrightarrow	⇎

This table is divided into three parts. The top part contains the symbols provided by LaTeX; the last command, \leadsto, requires the latexsym package. The middle table contains the $\mathcal{AMS}$ arrows; they all require the amssymb package. The bottom table lists the negated arrow symbols; they also require amssymb.

A.8   Miscellaneous symbols

Type:	Print:	Type:	Print:
\hbar	ℏ	\ell	ℓ
\imath	ı	\jmath	ȷ
\wp	℘	\Re	ℜ
\Im	ℑ	\partial	∂
\infty	∞	\prime	′
\emptyset	∅	\backslash	\
\forall	∀	\exists	∃
\smallint	∫	\triangle	△
\surd	√	\Vert	‖
\top	⊤	\bot	⊥
\P	¶	\S	§
\dag	†	\ddag	‡
\flat	♭	\natural	♮
\sharp	♯	\angle	∠
\clubsuit	♣	\diamondsuit	◇
\heartsuit	♡	\spadesuit	♠
\neg	¬		
\Box	□	\Diamond	◇
\mho	℧		
\hslash	ℏ	\complement	∁
\backprime	‵	\vartriangle	△
\Bbbk	𝕜	\varnothing	∅
\diagup	╱	\diagdown	╲
\blacktriangle	▲	\blacktriangledown	▼
\triangledown	▽	\Game	⅁
\square	□	\blacksquare	■
\lozenge	◇	\blacklozenge	◆
\measuredangle	∡	\sphericalangle	∢
\circledS	Ⓢ	\bigstar	★
\Finv	Ⅎ	\eth	ð
\nexists	∄		

This table is divided into two parts. The top part contains the symbols provided by LaTeX; the last three commands require the latexsym package. The bottom table lists symbols from the 𝒜ℳ𝒮; they all require the amssymb package.

A.9 *Math spacing commands*

Short form:	Full form:	Size:	Short form:	Full form:
\\,	\thinspace	⊔	\\!	\negthinspace
\\:	\medspace	⊔		\negmedspace
\\;	\thickspace	⊔		\negthickspace
	\quad	⊔		
	\qquad	⊔		

The \medspace, \thickspace, \negmedspace, and \negthickspace commands require the amsmath package.

A.10 *Delimiters*

Name:	Type:	Print:	Name:	Type:	Print:
Left paren	(	(	Right paren	)	)
Left bracket	[	[	Right bracket	]	]
Left brace	\\{	{	Right brace	\\}	}
Reverse slash	\backslash	\	Forward slash	/	/
Left angle	\langle	⟨	Right angle	\rangle	⟩
Vertical line	|	|	Double vert. line	\\|	‖
Left floor	\lfloor	⌊	Right floor	\rfloor	⌋
Left ceiling	\lceil	⌈	Right ceiling	\rceil	⌉
Upper left corner	\ulcorner	⌜	Upper right corner	\urcorner	⌝
Lower left corner	\llcorner	⌞	Lower right corner	\lrcorner	⌟

The corners require the amsmath package.

Name:	Type:	Print:
Upward arrow	\uparrow	↑
Double upward arrow	\Uparrow	⇑
Downward arrow	\downarrow	↓
Double downward arrow	\Downarrow	⇓
Up-and-down arrow	\updownarrow	↕
Double up-and-down arrow	\Updownarrow	⇕

A.11   *Operators*

\arccos	\arcsin	\arctan	\arg
\cos	\cosh	\cot	\coth
\csc	\dim	\exp	\hom
\ker	\lg	\ln	\log
\sec	\sin	\sinh	\tan
\tanh			
\varliminf	\varlimsup	\varinjlim	\varprojlim

The \var commands require the amsmath package.

\det	\gcd	\inf	\injlim
\lim	\liminf	\limsup	\max
\min	\projlim	\Pr	\sup

The \injlim and \projlim commands require the amsmath package.

Type:	Inline	Displayed	Type:	Inline	Displayed
\prod_{i=1}^{n}	$\prod_{i=1}^{n}$	$\displaystyle\prod_{i=1}^{n}$	\coprod_{i=1}^{n}	$\coprod_{i=1}^{n}$	$\displaystyle\coprod_{i=1}^{n}$
\bigcap_{i=1}^{n}	$\bigcap_{i=1}^{n}$	$\displaystyle\bigcap_{i=1}^{n}$	\bigcup_{i=1}^{n}	$\bigcup_{i=1}^{n}$	$\displaystyle\bigcup_{i=1}^{n}$
\bigwedge_{i=1}^{n}	$\bigwedge_{i=1}^{n}$	$\displaystyle\bigwedge_{i=1}^{n}$	\bigvee_{i=1}^{n}	$\bigvee_{i=1}^{n}$	$\displaystyle\bigvee_{i=1}^{n}$
\bigsqcup_{i=1}^{n}	$\bigsqcup_{i=1}^{n}$	$\displaystyle\bigsqcup_{i=1}^{n}$	\biguplus_{i=1}^{n}	$\biguplus_{i=1}^{n}$	$\displaystyle\biguplus_{i=1}^{n}$
\bigotimes_{i=1}^{n}	$\bigotimes_{i=1}^{n}$	$\displaystyle\bigotimes_{i=1}^{n}$	\bigoplus_{i=1}^{n}	$\bigoplus_{i=1}^{n}$	$\displaystyle\bigoplus_{i=1}^{n}$
\bigodot_{i=1}^{n}	$\bigodot_{i=1}^{n}$	$\displaystyle\bigodot_{i=1}^{n}$	\sum_{i=1}^{n}	$\sum_{i=1}^{n}$	$\displaystyle\sum_{i=1}^{n}$

A.12 Math accents

\hat{a}	$\hat{a}$	\Hat{a}	$\hat{a}$	\widehat{a}	$\widehat{a}$	a\sphat	$a^{\wedge}$
\tilde{a}	$\tilde{a}$	\Tilde{a}	$\tilde{a}$	\widetilde{a}	$\widetilde{a}$	a\sptilde	$a^{\sim}$
\acute{a}	$\acute{a}$	\Acute{a}	$\acute{a}$				
\bar{a}	$\bar{a}$	\Bar{a}	$\bar{a}$				
\breve{a}	$\breve{a}$	\Breve{a}	$\breve{a}$			a\spbreve	$a^{\breve{}}$
\check{a}	$\check{a}$	\Check{a}	$\check{a}$			a\spcheck	$a^{\vee}$
\dot{a}	$\dot{a}$	\Dot{a}	$\dot{a}$			a\spdot	$a^{\cdot}$
\ddot{a}	$\ddot{a}$	\Ddot{a}	$\ddot{a}$			a\spddot	$a^{\cdot\cdot}$
\dddot{a}	$\dddot{a}$					a\spdddot	$a^{\cdot\cdot\cdot}$
\ddddot{a}	$\ddddot{a}$						
\grave{a}	$\grave{a}$	\Grave{a}	$\grave{a}$	\imath	$\imath$		
\vec{a}	$\vec{a}$	\Vec{a}	$\vec{a}$	\jmath	$\jmath$		

The \dddot and \ddddot commands and all the capitalized commands require the amsmath package; the commands in the fourth column require the amsxtra package.

A.13 Math font commands

Type:	Print:
\mathbf{A}	$\mathbf{A}$
\mathit{A}	A
\mathsf{A}	A
\mathrm{A}	A
\mathtt{A}	$\mathtt{A}$
\mathnormal{A}	$\mathnormal{A}$
\mathbb{A}	$\mathbb{A}$
\mathfrak{A}	$\mathfrak{A}$
\mathcal{A}	$\mathcal{A}$
\boldsymbol{\alpha}	$\boldsymbol{\alpha}$

The \mathbb, \mathfrak, and \mathcal commands require the amsfonts package. The \boldsymbol command requires the amsbsy package.

Text symbol tables

B.1 Special text characters

Type:	Print:	Type:	Print:	Type:	Print:
\#	#	\$	$	\%	%
\&	&	\~{}	~	_	_
\^{}	^	\{	{	\}	}
$\|$	|	@	@	$*$	*
		$\backslash$	\		

B.2 Text accents

Type:	Print:	Type:	Print:	Type:	Print:
\'{o}	ò	\'{o}	ó	\"{o}	ö
\H{o}	ő	\^{o}	ô	\~{o}	õ
\v{o}	ǒ	\u{o}	ŏ	\={o}	ō
\b{o}	o̲	\.{o}	ȯ	\d{o}	ọ
\c{o}	ǫ	\r{o}	o̊	\t{oo}	o͡o
\i	ı			\j	ȷ

B.3 Some European characters

Type:	Print:	Type:	Print:	Type:	Print:
\aa	å	\AA	Å	\ae	æ
\AE	Æ	\o	ø	\O	Ø
\oe	œ	\OE	Œ	\l	ł
\L	Ł	\ss	ß	\SS	SS
?`	¿	!`	¡		

B.4 Extra text symbols

Type:	Print:
\dag	†
\ddag	‡
\S	§
\P	¶
\copyright	©
\pounds	£
\textbullet	•
\textvisiblespace	␣
\textcircled{a}	ⓐ
\textperiodcentered	·

B.5 Text spacing commands

Short form:	Full form:	Size:	Short form:	Full form:
\,	\thinspace	␣	\!	\negthinspace
\:	\medspace	␣		\negmedspace
\;	\thickspace	␣		\negthickspace
	\quad	⎵		
	\qquad	⎵⎵		

The \medspace, \thickspace, \negmedspace, and \negthickspace
commands require the amsmath package.

B.6 Text font commands

command with argument	command declaration	switch to
\textnormal{...}	{\normalfont ...}	document font family
\textrm{...}	{\rmfamily ...}	roman font family
\textsf{...}	{\sffamily ...}	sans serif font family
\texttt{...}	{\ttfamily ...}	typewriter style font family
\textup{...}	{\upshape ...}	upright shape
\textit{...}	{\itshape ...}	*italic shape*
\textsl{...}	{\slshape ...}	*slanted shape*
\textnormal{...}	{\normalfont ...}	default font
\textsc{...}	{\scshape ...}	SMALL CAPITALS
\emph{...}	{\em ...}	*emphasis*
\textbf{...}	{\bfseries ...}	**bold** (extended)
\textmd{...}	{\mdseries ...}	normal weight and width

B.7 *Text font size changes*

`\tiny`	<small>sample text</small>
`\scriptsize`	sample text
`\footnotesize`	sample text
`\small`	sample text
`\normalsize`	sample text
`\large`	sample text
`\Large`	sample text
`\LARGE`	sample text
`\huge`	sample text
`\Huge`	sample text

B.8 *A$_{\mathcal{M}}$S text font size changes*

```
\Tiny
\large
\tiny
\Large
\SMALL

\normalsize

\LARGE
\Small
\huge
\small
\Huge
```

The $\mathcal{A}\!\mathcal{M}\!\mathcal{S}$-L^{A}T$_{E}$X sample article

sampart.tex is the source file for our sample article using the $\mathcal{A}\!\mathcal{M}\!\mathcal{S}$-L^{A}T$_{E}$X article document class, amsart. A simpler article, intrart.tex, using the article document class was presented in Part I (see section 1.5).

The typeset sampart.tex is printed on the following three pages; it is followed by the source file and the typeset version shown together, so you can see how the marked up source file is turned into the typeset article.

A CONSTRUCTION OF COMPLETE-SIMPLE
DISTRIBUTIVE LATTICES

G. A. MENUHIN

ABSTRACT. In this note we prove that there exist *complete-simple distributive lattices*, that is, complete distributive lattices in which there are only two complete congruences.

1. INTRODUCTION

In this note we prove the following result:

Main Theorem. *There exists an infinite complete distributive lattice K with only the two trivial complete congruence relations.*

2. THE $D^{\langle 2 \rangle}$ CONSTRUCTION

For the basic notation in lattice theory and universal algebra, see F. R. Richardson [5] and G. A. Menuhin [2]. We start with some definitions:

Definition 1. Let V be a complete lattice, and let $\mathfrak{p} = [u, v]$ be an interval of V. Then $\mathfrak{p}$ is called *complete-prime* if the following three conditions are satisfied:

(1) u is meet-irreducible but u is *not* completely meet-irreducible;
(2) v is join-irreducible but v is *not* completely join-irreducible;
(3) $[u, v]$ is a complete-simple lattice.

Now we prove the following result:

Lemma 1. *Let D be a complete distributive lattice satisfying conditions (1) and (2). Then $D^{\langle 2 \rangle}$ is a sublattice of D^2; hence $D^{\langle 2 \rangle}$ is a lattice, and $D^{\langle 2 \rangle}$ is a complete distributive lattice satisfying conditions (1) and (2).*

Proof. By conditions (1) and (2), $D^{\langle 2 \rangle}$ is a sublattice of D^2. Hence, $D^{\langle 2 \rangle}$ is a lattice.

Since $D^{\langle 2 \rangle}$ is a sublattice of a distributive lattice, $D^{\langle 2 \rangle}$ is a distributive lattice. Using the characterization of standard ideals in Ernest T. Moynahan [3], $D^{\langle 2 \rangle}$ has a zero and a unit element, namely, $\langle 0, 0 \rangle$ and $\langle 1, 1 \rangle$. To show that $D^{\langle 2 \rangle}$ is complete, let $\varnothing \neq A \subseteq D^{\langle 2 \rangle}$, and let $a = \bigvee A$ in D^2. If $a \in D^{\langle 2 \rangle}$, then $a = \bigvee A$ in $D^{\langle 2 \rangle}$; otherwise, a is of the form $\langle b, 1 \rangle$ for some $b \in D$ with $b < 1$. Now $\bigvee A = \langle 1, 1 \rangle$ in D^2 and the dual argument shows that $\bigwedge A$ also exists in D^2. Hence D is complete. Conditions (1) and (2) are obvious for $D^{\langle 2 \rangle}$. $\qquad\square$

Corollary 1. *If D is complete-prime, then so is $D^{\langle 2 \rangle}$.*

Date: March 15, 1995.

1991 *Mathematics Subject Classification*. Primary: 06B10; Secondary: 06D05.

Key words and phrases. Complete lattice, distributive lattice, complete congruence, congruence lattice.

Research supported by the NSF under grant number 23466.

The motivation for the following result comes from S.-K. Foo [1].

Lemma 2. *Let Θ be a complete congruence relation of $D^{\langle 2 \rangle}$ such that*

$$(2.1) \qquad\qquad \langle 1, d \rangle \equiv \langle 1, 1 \rangle \pmod{\Theta},$$

for some $d \in D$ with $d < 1$. Then $\Theta = \iota$.

Proof. Let Θ be a complete congruence relation of $D^{\langle 2 \rangle}$ satisfying (2.1). Then $\Theta = \iota$.   □

3. THE Π^* CONSTRUCTION

The following construction is crucial to our proof of the Main Theorem:

Definition 2. Let D_i for $i \in I$ be complete distributive lattices satisfying condition (2). Their Π^* product is defined as follows:

$$\Pi^*(D_i \mid i \in I) = \Pi(D_i^- \mid i \in I) + 1;$$

that is, $\Pi^*(D_i \mid i \in I)$ is $\Pi(D_i^- \mid i \in I)$ with a new unit element.

Notation. If $i \in I$ and $d \in D_i^-$, then

$$\langle \ldots, 0, \ldots, \overset{i}{d}, \ldots, 0, \ldots \rangle$$

is the element of $\Pi^*(D_i \mid i \in I)$ whose ith component is d and all the other components are 0.

See also E. T. Moynahan [4]. Next we verify:

Theorem 1. *Let D_i for $i \in I$ be complete distributive lattices satisfying condition (2). Let Θ be a complete congruence relation on $\Pi^*(D_i \mid i \in I)$. If there exist $i \in I$ and $d \in D_i$ with $d < 1_i$ such that for all $d \leq c < 1_i$,*

$$(3.1) \qquad \langle \ldots, 0, \ldots, \overset{i}{d}, \ldots, 0, \ldots \rangle \equiv \langle \ldots, 0, \ldots, \overset{i}{c}, \ldots, 0, \ldots \rangle \pmod{\Theta},$$

then $\Theta = \iota$.

Proof. Since

$$(3.2) \qquad \langle \ldots, 0, \ldots, \overset{i}{d}, \ldots, 0, \ldots \rangle \equiv \langle \ldots, 0, \ldots, \overset{i}{c}, \ldots, 0, \ldots \rangle \pmod{\Theta},$$

and Θ is a complete congruence relation, it follows from condition (3) that

$$(3.3) \qquad \langle \ldots, \overset{i}{d}, \ldots, 0, \ldots \rangle \equiv$$

$$\bigvee (\langle \ldots, 0, \ldots, \overset{i}{c}, \ldots, 0, \ldots \rangle \mid d \leq c < 1) \equiv 1 \pmod{\Theta}.$$

Let $j \in I$ for $j \neq i$, and let $a \in D_j^-$. Meeting both sides of the congruence (3.2) with $\langle \ldots, 0, \ldots, \overset{j}{a}, \ldots, 0, \ldots \rangle$, we obtain

$$(3.4) \qquad 0 = \langle \ldots, 0, \ldots, \overset{i}{d}, \ldots, 0, \ldots \rangle \wedge \langle \ldots, 0, \ldots, \overset{j}{a}, \ldots, 0, \ldots \rangle \equiv$$

$$\langle \ldots, 0, \ldots, \overset{j}{a}, \ldots, 0, \ldots \rangle \pmod{\Theta}.$$

Using the completeness of Θ and (3.4), we get:

$$0 \equiv \bigvee (\langle \ldots, 0, \ldots, \overset{j}{a}, \ldots, 0, \ldots \rangle \mid a \in D_j^-) = 1 \pmod{\Theta},$$

hence $\Theta = \iota$.   □

hence $\Theta = \iota$. $\qquad\qquad\qquad\qquad\qquad\qquad\qquad\qquad\qquad\qquad$ □

Theorem 2. *Let D_i for $i \in I$ be complete distributive lattices satisfying conditions (2) and (3). Then $\Pi^*(D_i \mid i \in I)$ also satisfies conditions (2) and (3).*

Proof. Let Θ be a complete congruence on $\Pi^*(D_i \mid i \in I)$. Let $i \in I$. Define

$$\widehat{D}_i = \{\langle \ldots, 0, \ldots, \overset{i}{d}, \ldots, 0, \ldots \rangle \mid d \in D_i^-\} \cup \{1\}.$$

Then $\widehat{D}_i$ is a complete sublattice of $\Pi^*(D_i \mid i \in I)$, and $\widehat{D}_i$ is isomorphic to D_i. Let Θ_i be the restriction of Θ to $\widehat{D}_i$.

Since D_i is complete-simple, so is $\widehat{D}_i$, and hence Θ_i is ω or ι. If $\Theta_i = \rho$ for all $i \in I$, then $\Theta = \omega$. If there is an $i \in I$, such that $\Theta_i = \iota$, then $0 \equiv 1 \pmod{\Theta}$, hence $\Theta = \iota$. $\qquad\qquad\qquad\qquad\qquad\qquad\qquad\qquad\qquad$ □

The Main Theorem follows easily from Theorems 1 and 2.

References

[1] Soo-Key Foo, *Lattice Constructions*, Ph.D. thesis, University of Winnebago, Winnebago, MN, December 1990.

[2] George A. Menuhin, *Universal Algebra*, D. van Nostrand, Princeton-Toronto-London-Melbourne, 1968.

[3] Ernest T. Moynahan, *On a problem of M. H. Stone*, Acta Math. Acad.Sci. Hungar. **8** (1957), 455–460.

[4] _____, *Ideals and congruence relations in lattices. II*, Magyar Tud. Akad. Mat. Fiz. Oszt. Közl. **9** (1957), 417–434 (Hungarian).

[5] Ferenc R. Richardson, *General Lattice Theory*, Mir, Moscow, expanded and revised ed., 1982 (Russian).

COMPUTER SCIENCE DEPARTMENT, UNIVERSITY OF WINNEBAGO, WINNEBAGO, MINNESOTA 23714

E-mail address: menuhin@ccw.uwinnebago.edu

```
% Sample file: sampart.tex
% The sample article for the amsart document class
% Typeset with LaTeX format

\documentclass{amsart}
\usepackage{amssymb}

\theoremstyle{plain}
\newtheorem{theorem}{Theorem}
\newtheorem{corollary}{Corollary}
\newtheorem*{main}{Main~Theorem}
\newtheorem{lemma}{Lemma}
\newtheorem{proposition}{Proposition}

\theoremstyle{definition}
\newtheorem{definition}{Definition}

\theoremstyle{remark}
\newtheorem*{notation}{Notation}

\numberwithin{equation}{section}

\begin{document}
\title[Complete-simple distributive lattices]
      {A construction of complete-simple\\
       distributive lattices}
\author{G. A. Menuhin}
\address{Computer Science Department\\
         University of Winnebago\\
         Winnebago, Minnesota 23714}
\email{menuhin@ccw.uwinnebago.edu}
\thanks{Research supported by the NSF under grant number~23466.}
\keywords{Complete lattice, distributive lattice, complete congruence,
      congruence lattice}
\subjclass{Primary: 06B10; Secondary: 06D05}
\date{March 15, 1995}
\begin{abstract}
   In this note we prove that there exist \emph{complete-simple distributive
   lattices}, that is, complete distributive lattices in which there are
   only two complete congruences.
\end{abstract}
\maketitle

\section{Introduction} \label{S:intro}
In this note we prove the following result:

\begin{main}
   There exists an infinite complete distributive lattice $K$ with only
   the two trivial complete congruence relations.
\end{main}

\section{The $D^{\langle 2 \rangle}$ construction} \label{S:Ds}
For the basic notation in lattice theory and universal algebra, see F.~R.
Richardson~\cite{fR82} and G.~A. Menuhin~\cite{gM68}.  We start with some
definitions:
```

A CONSTRUCTION OF COMPLETE-SIMPLE
DISTRIBUTIVE LATTICES

G. A. MENUHIN

ABSTRACT. In this note we prove that there exist *complete-simple distributive lattices*, that is, complete distributive lattices in which there are only two complete congruences.

1. INTRODUCTION

In this note we prove the following result:

Main Theorem. *There exists an infinite complete distributive lattice K with only the two trivial complete congruence relations.*

2. THE $D^{\langle 2 \rangle}$ CONSTRUCTION

For the basic notation in lattice theory and universal algebra, see F. R. Richardson [5] and G. A. Menuhin [2]. We start with some definitions:

Date: March 15, 1995.

1991 *Mathematics Subject Classification*. Primary: 06B10; Secondary: 06D05.

Key words and phrases. Complete lattice, distributive lattice, complete congruence, congruence lattice.

Research supported by the NSF under grant number 23466.

Definition 1. Let V be a complete lattice, and let $\mathfrak{p} = [u, v]$ be an interval of V. Then $\mathfrak{p}$ is called *complete-prime* if the following three conditions are satisfied:

- (1) u is meet-irreducible but u is *not* completely meet-irreducible;
- (2) v is join-irreducible but v is *not* completely join-irreducible;
- (3) $[u, v]$ is a complete-simple lattice.

Now we prove the following result:

Lemma 1. *Let D be a complete distributive lattice satisfying conditions* (1) *and* (2). *Then $D^{\langle 2 \rangle}$ is a sublattice of D^2; hence $D^{\langle 2 \rangle}$ is a lattice, and $D^{\langle 2 \rangle}$ is a complete distributive lattice satisfying conditions* (1) *and* (2).

Proof. By conditions (1) and (2), $D^{\langle 2 \rangle}$ is a sublattice of D^2. Hence, $D^{\langle 2 \rangle}$ is a lattice.

```
\begin{definition} \label{D:prime}
    Let $V$ be a complete lattice, and let $\mathfrak{p} = [u, v]$ be an interval
    of $V$.  Then $\mathfrak{p}$ is called \emph{complete-prime} if the following
    three conditions are satisfied:
    \begin{itemize}
        \item[(1)] $u$ is meet-irreducible but $u$ is \emph{not}
            completely meet-irreducible;
        \item[(2)] $v$ is join-irreducible but $v$ is \emph{not}
            completely join-irreducible;
        \item[(3)] $[u, v]$ is a complete-simple lattice.
    \end{itemize}
\end{definition}

Now we prove the following result:
\begin{lemma} \label{L:ds}
    Let $D$ be a complete distributive lattice satisfying conditions~\textup{(1)}
    and \textup{(2)}.  Then $D^{\langle 2 \rangle}$ is a sublattice of $D^{2}$;
    hence $D^{\langle 2 \rangle}$ is a lattice, and $D^{\langle 2 \rangle}$
    is a complete distributive lattice satisfying conditions~\textup{(1)} and
    \textup{(2)}.
\end{lemma}
\begin{proof}
    By conditions~(1) and (2), $D^{\langle 2 \rangle}$ is a sublattice
    of $D^{2}$.  Hence, $D^{\langle 2 \rangle}$ is a lattice.
```

Since $D^{\langle 2 \rangle}$ is a sublattice of a distributive lattice, $D^{\langle 2 \rangle}$ is a distributive lattice. Using the characterization of standard ideals in Ernest T. Moynahan [3], $D^{\langle 2 \rangle}$ has a zero and a unit element, namely, $\langle 0, 0 \rangle$ and $\langle 1, 1 \rangle$. To show that $D^{\langle 2 \rangle}$ is complete, let $\varnothing \ne A \subseteq D^{\langle 2 \rangle}$, and let $a = \bigvee A$ in D^2. If $a \in D^{\langle 2 \rangle}$, then $a = \bigvee A$ in $D^{\langle 2 \rangle}$; otherwise, a is of the form $\langle b, 1 \rangle$ for some $b \in D$ with $b < 1$. Now $\bigvee A = \langle 1, 1 \rangle$ in D^2 and the dual argument shows that $\bigwedge A$ also exists in D^2. Hence D is complete. Conditions (1) and (2) are obvious for $D^{\langle 2 \rangle}$.   □

Corollary 1. *If D is complete-prime, then so is $D^{\langle 2 \rangle}$.*

The motivation for the following result comes from S.-K. Foo [1].

Lemma 2. *Let Θ be a complete congruence relation of $D^{\langle 2 \rangle}$ such that*

$$(2.1) \qquad \langle 1, d \rangle \equiv \langle 1, 1 \rangle \pmod{\Theta},$$

for some $d \in D$ with $d < 1$. Then $\Theta = \iota$.

```
Since $D^{\langle 2 \rangle}$ is a sublattice of a distributive lattice,
$D^{\langle 2 \rangle}$ is a distributive lattice.  Using the characterization
of standard ideals in Ernest~T. Moynahan~\cite{eM57}, $D^{\langle 2 \rangle}$
has a zero and a unit element, namely, $\langle 0, 0 \rangle$ and
$\langle 1, 1 \rangle$.  To show that $D^{\langle 2 \rangle}$ is complete,
let $\varnothing \ne A \subseteq D^{\langle 2 \rangle}$, and let
$a = \bigvee A$ in $D^{2}$.  If $a \in D^{\langle 2 \rangle}$, then
$a = \bigvee A$ in $D^{\langle 2 \rangle}$; otherwise, $a$ is of the form
$\langle b, 1 \rangle$ for some $b \in D$ with $b < 1$.  Now
$\bigvee A = \langle 1, 1\rangle$ in $D^{2}$ and the dual argument shows that
$\bigwedge A$ also exists in $D^{2}$.  Hence $D$ is complete. Conditions~(1)
and (2) are obvious for $D^{\langle 2 \rangle}$.
\end{proof}
\begin{corollary} \label{C:prime}
   If $D$ is complete-prime, then so is $D^{\langle 2 \rangle}$.
\end{corollary}

The motivation for the following result comes from S.-K. Foo~\cite{sF90}.
\begin{lemma} \label{L:ccr}
   Let $\Theta$ be a complete congruence relation of
   $D^{\langle 2 \rangle}$ such that
   \begin{equation} \label{E:rigid}
      \langle 1, d \rangle \equiv \langle 1, 1 \rangle \pmod{\Theta},
   \end{equation}
   for some $d \in D$ with $d < 1$. Then $\Theta = \iota$.
\end{lemma}
```

Proof. Let Θ be a complete congruence relation of $D^{\langle 2 \rangle}$ satisfying (2.1). Then $\Theta = \iota$. $\square$

3. THE Π^* CONSTRUCTION

The following construction is crucial to our proof of the Main Theorem:

Definition 2. Let D_i for $i \in I$ be complete distributive lattices satisfying condition (2). Their Π^* product is defined as follows:

$$\Pi^*(D_i \mid i \in I) = \Pi(D_i^- \mid i \in I) + 1;$$

that is, $\Pi^*(D_i \mid i \in I)$ is $\Pi(D_i^- \mid i \in I)$ with a new unit element.

Notation. If $i \in I$ and $d \in D_i^-$, then

$$\langle \dots, 0, \dots, \overset{i}{d}, \dots, 0, \dots \rangle$$

is the element of $\Pi^*(D_i \mid i \in I)$ whose ith component is d and all the other components are 0.

```
\begin{proof}
   Let $\Theta$ be a complete congruence relation of $D^{\langle 2 \rangle}$
   satisfying \eqref{E:rigid}. Then $\Theta = \iota$.
\end{proof}

\section{The $\Pi^{*}$ construction} \label{S:P*}
The following construction is crucial to our proof of the Main Theorem:
\begin{definition} \label{D:P*}
   Let $D_{i}$ for $i \in I$ be complete distributive lattices satisfying
   condition~\textup{(2)}.  Their $\Pi^{*}$ product is defined as follows:
   \[
      \Pi^{*} ( D_{i} \mid i \in I ) = \Pi ( D_{i}^{-} \mid i \in I ) + 1;
   \]
   that is, $\Pi^{*} ( D_{i} \mid i \in I )$ is $\Pi ( D_{i}^{-} \mid
   i \in I )$ with a new unit element.
\end{definition}
\begin{notation}
   If $i \in I$ and $d \in D_{i}^{-}$, then
   \[
      \langle \dots, 0, \dots, \overset{i}{d}, \dots, 0, \dots \rangle
   \]
   is the element of $\Pi^{*} ( D_{i} \mid i \in I )$ whose $i$th
   component is $d$ and all the other components are $0$.
\end{notation}
```

See also E. T. Moynahan [4]. Next we verify:

Theorem 1. *Let D_i for $i \in I$ be complete distributive lattices satisfying condition* (2). *Let Θ be a complete congruence relation on $\Pi^*(D_i \mid i \in I)$. If there exist $i \in I$ and $d \in D_i$ with $d < 1_i$ such that for all $d \leq c < 1_i$,*

$$(3.1) \qquad \langle \dots, 0, \dots, \overset{i}{d}, \dots, 0, \dots \rangle \equiv \langle \dots, 0, \dots, \overset{i}{c}, \dots, 0, \dots \rangle \pmod{\Theta},$$

then $\Theta = \iota$.

Proof. Since

$$(3.2) \qquad \langle \dots, 0, \dots, \overset{i}{d}, \dots, 0, \dots \rangle \equiv \langle \dots, 0, \dots, \overset{i}{c}, \dots, 0, \dots \rangle \pmod{\Theta},$$

and Θ is a complete congruence relation, it follows from condition (3) that

$$(3.3) \qquad \langle \dots, \overset{i}{d}, \dots, 0, \dots \rangle \equiv$$
$$\bigvee (\langle \dots, 0, \dots, \overset{i}{c}, \dots, 0, \dots \rangle \mid d \leq c < 1) \equiv 1 \pmod{\Theta}.$$

Let $j \in I$ for $j \neq i$, and let $a \in D_j^-$. Meeting both sides of the congruence (3.2) with $\langle \dots, 0, \dots, \overset{j}{a}, \dots, 0, \dots \rangle$, we obtain

```
See also E.~T. Moynahan \cite{eM57a}.  Next we verify:
\begin{theorem} \label{T:P*}
   Let $D_{i}$ for $i \in I$ be complete distributive lattices satisfying
   condition~\textup{(2)}.  Let $\Theta$ be a complete congruence relation
   on $\Pi^{*} ( D_{i} \mid i \in I )$.  If there exist $i \in I$
   and $d \in D_{i}$ with $d < 1_{i}$ such that for all $d \leq c < 1_{i}$,
   \begin{equation} \label{E:cong1}
      \langle \dots, 0, \dots,\overset{i}{d}, \dots, 0, \dots \rangle \equiv
      \langle \dots, 0, \dots, \overset{i}{c}, \dots, 0, \dots \rangle \pmod{\Theta},
   \end{equation}
   then $\Theta = \iota$.
\end{theorem}
\begin{proof}
   Since
   \begin{equation} \label{E:cong2}
      \langle \dots, 0, \dots, \overset{i}{d}, \dots, 0, \dots \rangle \equiv
      \langle \dots, 0, \dots, \overset{i}{c}, \dots, 0, \dots \rangle \pmod{\Theta},
   \end{equation}
   and $\Theta$ is a complete congruence relation, it follows from
   condition~(3) that
   \begin{align} \label{E:cong}
      & \langle \dots, \overset{i}{d}, \dots, 0, \dots \rangle \equiv\\
      &\qquad \quad \bigvee ( \langle \dots, 0, \dots,
      \overset{i}{c}, \dots, 0, \dots \rangle \mid d \leq c < 1 )
      \equiv 1 \pmod{\Theta}. \notag
   \end{align}
   Let $j \in I$ for $j \neq i$, and let $a \in D_{j}^{-}$.  Meeting both
   sides of the congruence \eqref{E:cong2} with $\langle \dots, 0, \dots,
   \overset{j}{a}, \dots, 0, \dots \rangle$, we obtain
```

(3.4) $0 = \langle \dots, 0, \dots, \overset{i}{d}, \dots, 0, \dots \rangle \wedge \langle \dots, 0, \dots, \overset{j}{a}, \dots, 0, \dots \rangle \equiv$

$$\langle \dots, 0, \dots, \overset{j}{a}, \dots, 0, \dots \rangle \pmod{\Theta}.$$

Using the completeness of Θ and (3.4), we get:

$$0 \equiv \bigvee (\langle \dots, 0, \dots, \overset{j}{a}, \dots, 0, \dots \rangle \mid a \in D_j^-) = 1 \pmod{\Theta},$$

hence $\Theta = \iota$.   □

Theorem 2. *Let D_i for $i \in I$ be complete distributive lattices satisfying conditions (2) and (3). Then $\Pi^*(D_i \mid i \in I)$ also satisfies conditions (2) and (3).*

Proof. Let Θ be a complete congruence on $\Pi^*(D_i \mid i \in I)$. Let $i \in I$. Define

$$\widehat{D}_i = \{ \langle \dots, 0, \dots, \overset{i}{d}, \dots, 0, \dots \rangle \mid d \in D_i^- \} \cup \{1\}.$$

Then $\widehat{D}_i$ is a complete sublattice of $\Pi^*(D_i \mid i \in I)$, and $\widehat{D}_i$ is isomorphic to D_i. Let Θ_i be the restriction of Θ to $\widehat{D}_i$.

```
\begin{align} \label{E:comp}
   0 = & \langle \dots, 0, \dots, \overset{i}{d}, \dots, 0, \dots
        \rangle \wedge \langle \dots, 0, \dots, \overset{j}{a}, \dots,
        0, \dots \rangle \equiv\\
        &\langle \dots, 0, \dots, \overset{j}{a}, \dots, 0, \dots
        \rangle \pmod{\Theta}. \notag
\end{align}
Using the completeness of $\Theta$ and \eqref{E:comp}, we get:
\[
   0 \equiv \bigvee ( \langle \dots, 0, \dots, \overset{j}{a},
   \dots, 0, \dots \rangle \mid a \in D_{j}^{-} ) = 1 \pmod{\Theta},
\]
hence $\Theta = \iota$.
\end{proof}
\begin{theorem} \label{T:P*a}
   Let $D_{i}$ for $i \in I$ be complete distributive lattices satisfying
   conditions \textup{(2)} and \textup{(3)}.  Then
   $\Pi^{*} ( D_{i} \mid i \in I )$
   also satisfies conditions \textup{(2)} and \textup{(3)}.
\end{theorem}
\begin{proof}
   Let $\Theta$ be a complete congruence on
   $\Pi^{*} ( D_{i} \mid i \in I )$. Let $i \in I$.  Define
   \[
      \widehat{D}_{i} = \{ \langle \dots, 0, \dots, \overset{i}{d},
      \dots, 0, \dots \rangle \mid d \in D_{i}^{-} \} \cup \{ 1 \}.
   \]
   Then $\widehat{D}_{i}$ is a complete sublattice of $\Pi^{*} ( D_{i} \
   mid i \in I )$, and $\widehat{D}_{i}$ is isomorphic to $D_{i}$.  Let
   $\Theta_{i}$ be the restriction of $\Theta$ to $\widehat{D}_{i}$.
```

Since D_i is complete-simple, so is $\widehat{D}_i$, and hence Θ_i is ω or ι. If $\Theta_i = \rho$ for all $i \in I$, then $\Theta = \omega$. If there is an $i \in I$, such that $\Theta_i = \iota$, then $0 \equiv 1 \pmod{\Theta}$, hence $\Theta = \iota$. $\qquad\qquad\qquad\qquad\qquad\qquad\qquad\qquad\qquad\qquad\qquad\qquad$ □

The Main Theorem follows easily from Theorems 1 and 2.

REFERENCES

[1] Soo-Key Foo, *Lattice Constructions*, Ph.D. thesis, University of Winnebago, Winnebago, MN, December 1990.

[2] George A. Menuhin, *Universal Algebra*, D. van Nostrand, Princeton-Toronto-London-Melbourne, 1968.

[3] Ernest T. Moynahan, *On a problem of M. H. Stone*, Acta Math. Acad.Sci. Hungar. **8** (1957), 455–460.

[4] ———, *Ideals and congruence relations in lattices. II*, Magyar Tud. Akad. Mat. Fiz. Oszt. Közl. **9** (1957), 417–434 (Hungarian).

[5] Ferenc R. Richardson, *General Lattice Theory*, Mir, Moscow, expanded and revised ed., 1982 (Russian).

COMPUTER SCIENCE DEPARTMENT, UNIVERSITY OF WINNEBAGO, WINNEBAGO, MINNESOTA 23714

E-mail address: `menuhin@ccw.uwinnebago.edu`

```
    Since $D_{i}\) is complete-simple, so is $\widehat{D}_{i}$, and hence
    $\Theta_{i}$ is $\omega$ or $\iota$.  If $\Theta_{i} = \rho$  for all
    $i \in I$, then $\Theta = \omega$.  If there is an $i \in I$, such that
    $\Theta_{i} = \iota$, then $0 \equiv 1 \pmod{\Theta}$, hence $\Theta = \iota$.
\end{proof}
The Main Theorem follows easily from Theorems~\ref{T:P*} and \ref{T:P*a}.
\begin{thebibliography}{9}
    \bibitem{sF90}
        Soo-Key Foo, \emph{Lattice Constructions}, Ph.D. thesis, University
        of Winnebago, Winnebago, MN, December 1990.
    \bibitem{gM68}
        George~A. Menuhin, \emph{Universal Algebra}, D.~van Nostrand,
        Princeton-Toronto-London-Mel\-bourne, 1968.
    \bibitem{eM57}
        Ernest~T. Moynahan, \emph{On a problem of M.~H. Stone}, Acta Math.
         Acad.Sci. Hungar. \textbf{8} (1957), 455--460.
    \bibitem{eM57a}
         \bysame, \emph{Ideals and congruence relations in lattices.~II},
        Magyar Tud. Akad. Mat. Fiz. Oszt. K\"{o}zl. \textbf{9} (1957),
        417--434  (Hungarian).
    \bibitem{fR82}
        Ferenc~R. Richardson, \emph{General Lattice Theory}, Mir, Moscow,
        expanded and revised ed., 1982 (Russian).
\end{thebibliography}
\end{document}
```

APPENDIX

D

The sample article with user-defined commands

In this appendix, we present the `sampart2.tex` sample article (also in the `ftp` directory), which is a rewrite of the `sampart.tex` sample article (see Appendix C and the `ftp` directory) utilizing the user-defined commands collected in `lattice.sty` (see section 9.5 and the `ftp` directory):

```
% Sample file: sampart2.tex
% The sample article for the amsart document class
% with user-defined commands
% Typeset with LaTeX format

\documentclass{amsart}
\usepackage{lattice}
\usepackage[notcite]{showkeys}% comment out for final version
\usepackage{xspace}

\theoremstyle{plain}
\newtheorem{theorem}{Theorem}
\newtheorem{corollary}{Corollary}
\newtheorem*{main}{Main~Theorem}
```

```
\newtheorem{lemma}{Lemma}
\newtheorem{proposition}{Proposition}

\theoremstyle{definition}
\newtheorem{definition}{Definition}

\theoremstyle{remark}
\newtheorem*{notation}{Notation}

\numberwithin{equation}{section}

\newcommand{\JJm}[2]{\JJ(\,#1\mid#2\,)}
   % big join with middle used as: \JJm{a}{a < 2}
\newcommand{\Prodm}[2]{\gP(\,#1\mid#2\,)}
   % product with a middle
\newcommand{\Prodsm}[2]{\gP^{*}(\,#1\mid#2\,)}
   % product * with a middle
\newcommand{\vct}[2]{\vv<\dots,0,\dots,\overset{#1}{#2},%
\dots,0,\dots>}% special vector
\newcommand{\fp}{\ensuremath{\F{p}}\xspace}
   % Fraktur p in text or math
\newcommand{\Ds}{\ensuremath{D^{\langle2\rangle}}\xspace}
   % \Ds in text or math


\begin{document}
\title[Complete-simple distributive lattices]
     {A construction of complete-simple\\
      distributive lattices}
\author{George~A. Menuhin}
\address{Computer Science Department\\
        University of Winnebago\\
        Winnebago, Minnesota 23714}
\email{menuhin@ccw.uwinnebago.edu}
\thanks{Research supported by the NSF under grant number~23466.}
\keywords{Complete lattice, distributive lattice, complete
   congruence, congruence lattice}
\subjclass{Primary: 06B10; Secondary: 06D05}
\date{March 15, 1995}


\begin{abstract}
   In this note we prove that there exist \emph{complete-simple
   distributive lattices}, that is, complete distributive
```

lattices in which there are only two complete congruences.
\end{abstract}
\maketitle

\section{Introduction}\label{S:intro}
In this note we prove the following result:

\begin{main}
    There exists an infinite complete distributive lattice
    K with only the two trivial complete congruence relations.
\end{main}

\section{The \Ds construction}\label{S:Ds}
For the basic notation in lattice theory and universal algebra,
see Ferenc~R. Richardson~\cite{fR82} and George~A.
Menuhin~\cite{gM68}. We start with some definitions:

\begin{definition}\label{D:prime}
    Let V be a complete lattice, and let $\fp = [u, v]$
    be an interval of V. Then \fp is called
    \emph{complete-prime} if the following three
    conditions are satisfied:
\begin{definition} \label{D:prime}
    Let V be a complete lattice, and let $\mathfrak{p} = [u, v]$
    be an interval of V. Then $\mathfrak{p}$ is called
    \emph{complete-prime} if the following
    three conditions are satisfied:
    \begin{itemize}
      \item[(1)] u is meet-irreducible but u is \emph{not}
         completely meet-irreducible;
      \item[(2)] v is join-irreducible but v is \emph{not}
         completely join-irreducible;
      \item[(3)] $[u, v]$ is a complete-simple lattice.
    \end{itemize}
\end{definition}

Now we prove the following result:

\begin{lemma}\label{L:ds}
    Let D be a complete distributive lattice satisfying
    conditions~\tup{(1)} and \tup{(2)}. Then \Ds is a
    sublattice of D^{2}; hence \Ds is a lattice, and

```
    \Ds is a complete distributive lattice satisfying
    conditions~\tup{(1)} and \tup{(2)}.
\end{lemma}

\begin{proof}
    By conditions~(1) and (2), \Ds is a sublattice of
    $D^{2}$.  Hence, \Ds is a lattice.

    Since \Ds is a sublattice of a distributive lattice, \Ds is
    a distributive lattice.  Using the characterization of
    standard ideals in E.~T. Moynahan~\cite{eM57},
    \Ds has a zero and a unit element, namely,
    $\vv<0, 0>$ and $\vv<1, 1>$.  To show that \Ds is
    complete, let $\es \ne A \ci \Ds$, and let $a = \JJ A$
    in $D^{2}$.  If $a \in \Ds$, then
    $a = \JJ A$ in \Ds; otherwise, $a$ is of the form
    $\vv<b, 1>$ for some $b \in D$ with $b < 1$.  Now
    $\JJ A = \vv<1, 1>$ in $D^{2}$, and
    the dual argument shows that $\MM A$ also exists in
    $D^{2}$.  Hence $D$ is complete. Conditions~(1) and (2)
    are obvious for \Ds.
\end{proof}

\begin{corollary}\label{C:prime}
    If $D$ is complete-prime, then so is \Ds.
\end{corollary}

The motivation for the following result comes from S.-K.
Foo~\cite{sF90}.

\begin{lemma}\label{L:ccr}
    Let $\gQ$ be a complete congruence relation of \Ds such
    that
    \begin{equation}\label{E:rigid}
        \con{\vv<1, d>}={\vv<1, 1>}(\gQ),
    \end{equation}
    for some $d \in D$ with $d < 1$. Then $\gQ = \gi$.
\end{lemma}

\begin{proof}
    Let $\gQ$ be a complete congruence relation of \Ds
    satisfying \eqref{E:rigid}. Then $\gQ = \gi$.
```

```
\end{proof}

\section{The $\gP^{*}$ construction}\label{S:P*}
The following construction is crucial to our proof of the
Main~Theorem:

\begin{definition}\label{D:P*}
   Let $D_{i}$ for $i \in I$ be complete distributive
   lattices satisfying condition~\tup{(2)}.  Their $\gP^{*}$
   product is defined as follows:
   \[
      \Prodsm{ D_{i} }{i \in I} = \Prodm{ D_{i}^{-} }{i \in I} +1;
   \]
   that is, $\Prodsm{ D_{i} }{i \in I}$ is
   $\Prodm{ D_{i}^{-} }{i \in I}$ with a new unit element.
\end{definition}

\begin{notation}
   If $i \in I$ and $d \in D_{i}^{-}$, then
   \[
      \vct{i}{d}
   \]
   is the element of $\Prodsm{ D_{i} }{i \in I}$ whose
   $i$th component is $d$ and all the other
   components are $0$.
\end{notation}

See also E.~T. Moynahan \cite{eM57a}.  Next we verify:

\begin{theorem}\label{T:P*}
   Let $D_{i}$ for $i \in I$ be complete distributive
   lattices satisfying condition~\tup{(2)}.  Let $\gQ$ be a
   complete congruence relation on
   $\Prodsm{ D_{i} }{i \in I}$.  If there exist
   $i \in I$ and $d \in D_{i}$ with $d < 1_{i}$ such
   that for all $d \leq c < 1_{i}$,
   \begin{equation}\label{E:cong1}
      \con\vct{i}{d}=\vct{i}{c}(\gQ),
   \end{equation}
   then $\gQ = \gi$.
\end{theorem}
```

```
\begin{proof}
   Since
   \begin{equation}\label{E:cong2}
      \con\vct{i}{d}=\vct{i}{c}(\gQ),
   \end{equation}
   and $\gQ$ is a complete congruence relation, it follows
   from condition~(3) that
   \begin{align}\label{E:cong}
      &\con{\vct{i}{d}}=\notag\\
      &\qq\q{\JJm{\vct{i}{c}}{d \leq c < 1}=1}(\gQ).
   \end{align}
   Let $j \in I$ for $j \neq i$, and let
   $a \in D_{j}^{-}\). Meeting both sides of the congruence
   \eqref{E:cong} with $\vct{j}{a}$, we obtain
   \begin{align}\label{E:comp}
      0 = &\vct{i}{d} \mm \vct{j}{a} \equiv\\
          &\vct{j}{a}\pod{\gQ}. \notag
   \end{align}
   Using the completeness of $\gQ$ and \eqref{E:comp}, we get:
   \begin{equation}\label{E:cong3}
      \con 0=\JJm{ \vct{j}{a} }{ a \in D_{j}^{-} }=1(\gQ),
   \end{equation}
   hence $\gQ = \gi$.
\end{proof}

\begin{theorem}\label{T:P*a}
   Let $D_{i}$ for $i \in I$ be complete distributive
   lattices satisfying
   conditions~\tup{(2)} and \tup{(3)}. Then
   $\Prodsm{ D_{i} }{i \in I}$ also satisfies
   conditions~\tup{(2)} and \tup{(3)}.
\end{theorem}

\begin{proof}
   Let $\gQ$ be a complete congruence on
   $\Prodsm{ D_{i} }{i \in I}$. Let $i \in I$. Define
   \begin{equation}\label{E:dihat}
      \widehat{D}_{i} = \setm{ \vct{i}{d} }{ d \in D_{i}^{-} }
      \uu \set{1}.
   \end{equation}
   Then $\widehat{D}_{i}$ is a complete sublattice of
   $\Prodsm{ D_{i} }{i \in I}$, and $\widehat{D}_{i}$
```

```
      is isomorphic to $D_{i}$.  Let $\gQ_{i}$ be the
      restriction of $\gQ$ to $\widehat{D}_{i}$.  Since
      $D_{i}$ is complete-simple, so is $\widehat{D}_{i}$,
      hence $\gQ_{i}$ is $\go$ or $\gi$.  If $\gQ_{i} = \go$
      for all $i \in I$, then $\gQ = \go$.
      If there is an $i \in I$, such that $\gQ_{i} = \gi$,
      then $\con 0=1(\gQ)$, and hence $\gQ = \gi$.
\end{proof}

The Main Theorem follows easily from Theorems~\ref{T:P*} and
\ref{T:P*a}.

\bibliographystyle{amsplain}
\bibliography{sampart1}

\end{document}
```

The showkeys package was discussed in section 6.4.2, while xspace in section 9.1.1. Note that the bib file of sampart1 (see section 10.2.1) is used here again.

E

Background

You need not know the genealogy and structure of LaTeX in order to work with it. However, I'll briefly outline these topics for the curious reader. This knowledge may help you to better understand the behavior of LaTeX.

E.1   A short history

Donald E. Knuth's multivolume work *The Art of Computer Programming* [23] caused a great deal of frustration to its author, since it was very difficult to keep the various volumes typographically uniform. To solve this problem, Knuth decided to create his own typesetting language (see [24]–[28]).[1]

A mathematical typesetting language takes care of the multitude of details that are so important in mathematical typesetting; it

- properly spaces the formulas;
- breaks up the text into pleasingly typeset lines and paragraphs;
- hyphenates words as necessary;

---

[1] In Software Practice and Experience **19** (1989), 607–685, Knuth writes that "[I] realized that a central aspect of printing has been reduced to bit manipulation. As a computer scientist, I could not resist the challenge of improving print quality by manipulating bits better."

- provides hundreds of symbols for typesetting mathematics.

TEX does all this and more on most any computer: IBM and IBM compatibles, Macintosh, Atari, Amiga, workstations, minicomputers, and mainframes. You can typeset your work on an IBM compatible and e-mail it to a coworker who'll do the corrections on a Macintosh, while the final result is e-mailed to your publisher who uses a minicomputer to print the document on a Linotype typesetter.

Knuth realized that typesetting is only half the solution to manuscript production. You also need a style designer—a specialist who decides what fonts to use, how large a vertical space to put after a theorem, and numerous other design issues that constitute a style. TEX was designed to work with a "document class", so you do not have to worry about style design problems.

Knuth also realized that typesetting a complex document in TEX requires a knowledgeable user. So TEX was designed as a "platform" on which *convenient work environments*—so called "macro packages"—could be built. It is somewhat unfortunate that *two* such macro packages were made available to the mathematical community in the early eighties: $\mathcal{AMS}$-TEX and LATEX.

$\mathcal{AMS}$-TEX was written by Michael D. Spivak for the $\mathcal{AMS}$, while LATEX was developed by Leslie Lamport. The strengths of the two systems are somewhat complementary. $\mathcal{AMS}$-TEX provided many features necessary for mathematical articles, including:

- Extensive options for formatting of aligned and other multiline formulas.
- Flexible bibliographic references.

LATEX also provided many features including:

- The use of logical units to separate the logical and the visual design of an article.
- Automatic numbering and cross-referencing.
- Bibliographic databases.

Both $\mathcal{AMS}$-TEX and LATEX became very popular, causing a split in the mathematical community.

Since Lamport decided not to develop LATEX any further, a talented group of mathematicians and programmers, Frank Mittelbach, Chris Rowley, and Rainer Schöpf, formed the "LATEX3 team" with the aim of updating, actively supporting, and maintaining LATEX. The group has since expanded with the addition of Johannes Braams, David Carlisle, Michael Downes, Denis Duchier, and Alan Jeffrey; many volunteers have also contributed time to the project.

The goals of LATEX3 are very ambitious:

- The LATEX3 system will provide high quality typesetting for a wide variety of document types and typographic requirements.

- For editors and designers, it'll support direct formatting commands, which are essential to the fine-tuning of document layout and page design.
- It'll process complex structured documents and support a document syntax that allows automatic translation of documents conforming to the international document type definition standard SGML (Standard Generalized Markup Language, ISO standard 8879).
- LaTeX3 will provide a common foundation for a number of incompatible LaTeX variants that have developed, including LaTeX with NFSS (New Font Selection Scheme), SLITEX (for slides), and AMS-LaTeX.

See Mittelbach and Rowley [33, 35] for a complete statement of goals and a progress report, respectively.

A number of projects have already been completed that will be part of LaTeX3 (in functionality, if not in code), including the following:

The New Font Selection Scheme. LaTeX uses Knuth's Computer Modern fonts. In 1989, Mittelbach and Schöpf coded NFSS that allows the *independent changing* of font attributes and the easy integration of new font families into LaTeX. With the proliferation of PostScript fonts and PostScript printers, more and more users want to use PostScript fonts.

AMS-TEX as a LaTeX option. Mittelbach and Schöpf (with Downes) recoded AMS-TEX so that it would work as a LaTeX option (see section E.1.1).

Proclamations with style. All proclamations in LaTeX were typeset in the same style, whether they were a **Main Theorem** or a lowly *Comment*. Mittelbach and Schöpf coded a sophisticated scheme that allowed proclamation styles to be specified.

New and improved environments. There are improved `verbatim` and `comment` environments by Schöpf, and there is a new multicolumn environment by Mittelbach. There are also several improvements to the `tabular` and `array` environments.

E.1.1   *The first interim solution*

In 1990, the AMS released AMS-LaTeX. This release contained AMS-TEX recoded as a LaTeX option, the NFSS, styles for proclamations, and the new `verbatim` environment.

While the LaTeX3 team wanted to unify the mathematical community, this first attempt split it even further. Many AMS-TEX users simply did not switch. Even the LaTeX community was split into users of the old LaTeX, those whose LaTeX incorporated the NFSS, and the new AMS-LaTeX users.

E.1.2   *The second interim solution*

When it became obvious that the goals of LaTeX3 could not be fulfilled any time soon, the LaTeX3 team decided to issue in June of 1994 a new standard version of LaTeX, version 2e (also called LaTeX 2$_\varepsilon$), to replace LaTeX version 2.09; see Mittelbach and Rowley [34, 36]. Some of the goals of LaTeX3 were accomplished with this interim release, including the projects listed above. As I am writing this, not quite a year has passed since the release of LaTeX 2$_\varepsilon$, but LaTeX 2$_\varepsilon$ has already established itself as standard LaTeX.

In February of 1995, the $\mathcal{AMS}$ released a new version of $\mathcal{AMS}$-LaTeX (version 1.2) built on the new LaTeX. Michael Downes was the project leader. The changes in $\mathcal{AMS}$-LaTeX were substantial. The `align` environment, for example, was completely rewritten by David M. Jones. The recoded $\mathcal{AMS}$-TeX is now a LaTeX package, amsmath.

It is really important to note that while $\mathcal{AMS}$-LaTeX version 1.1 was a monolithic structure, version 1.2 is just a collection of packages, that fit nicely into the LaTeX mindset. You can use one $\mathcal{AMS}$-LaTeX package or all, by themselves or intermixed with other LaTeX packages. This book uses a LaTeX document class and intermixes the $\mathcal{AMS}$-LaTeX packages with a number of LaTeX (non-$\mathcal{AMS}$) packages.

E.2   *How does it work?*

In this section, I present a very simplified overview of the inner-workings of LaTeX.

E.2.1   *The layers*

TeX and LaTeX have many layers. These include:

`virtex`

The core of TeX, called `virtex`, contains only the most primitive commands. It knows about 300 basic commands such as `\input`, `\accent`, and `\hsize` and has the ability to read in *format files*, which are "precompiled" sets of macros. Basically, LaTeX is `virtex` reading in a large set of macros, built layer upon layer.

`plain.tex`

`plain.tex`, the most basic layer on top of `virtex`, was created by Knuth. It adds about 600 commands to `virtex`. When you issue a `tex` command, it really executes `virtex` with the `plain` format, the default format.

`plain.tex` is described in detail in Appendix B of Knuth [24]. You can read `plain.tex` if you like: it's a text file in the TeX distribution. As a matter of fact, `plain.tex` is powerful enough so that you could do all your work in it. This view is advocated by many, for instance, by Michael Doob [10].

virtex can't build format files. For that you need another version of TEX, called initex. This calls in the most basic information, such as the hyphenation tables and plain.tex, and creates a format.

*L*A*TEX*

LATEX is a set of macros written by Lamport et al.; see the latex.ltx file. It provides for logical document design, automatic numbering and cross-referencing, tables of contents, and many other features. LATEX contains, with some modifications, the macros of plain.tex.

Document classes

The document class and its options form the next layer. You may choose to use a standard LATEX document class: report, article, book, proc, letter, or slides; one provided by the *AMS*-LATEX, including amsart, amsbook, or amsproc; or any one of a very large number of other document classes provided by publishers of books and journals, and other interested parties.

The packages

At the top of this hierarchy of layers are the packages. You can use standard LATEX packages, *AMS*-LATEX packages, or any one of hundreds of other packages in the LATEX universe, intermixed as necessary. A package may require another package, or in fact may load a number of other packages.

Whenever you use an *AMS*-LATEX document class, a number of packages are automatically loaded; see Figure 8.3.

E.2.2   Typesetting

When typesetting, TEX uses two basic types of files: the source file(s) and the font metric file(s).

There are font metric files for each font used (including each design size). Each TEX font metric file, called a tfm file, contains the size of each character, the measurements for "kerning" (the space placed between two adjacent characters), the length of the "italic correction", the size of the "interword space", and so on. A typical tfm file is cmr10.tfm, which is the TEX font metric file for the font cmr at size 10 points.

TEX reads the source file a line at a time. It converts the characters of each line into a token sequence: a "token" is either a character (together with an indication of what role the character plays) or a macro. The (undelimited) argument of a macro is the token following the macro, unless a group enclosed in braces follows the macro (in which case the contents of the group becomes the argument).

Similarly, when you exponentiate, TEX looks for the next token as the exponent unless a group enclosed in braces follows the ^ symbol. This may help explain why 2^3 and 2^α work out well, but $2^\mathfrak{m}$ does not: 3 and \alpha turn into a single token each, \mathfrak{m} turns into more than one (in fact, four) tokens. Of course, if you *always* use braces

```
$ 2^{3}$, $2^{\alpha}$, $2^{ \mathfrak{m} }$
```

then you do not have to remember what tokens are.

After tokenizing the text, TEX hyphenates it and attempts to split the paragraph into lines of the required width. The measurements of the characters are absolute, and so are the distances between characters (kerning); however, the spaces (interword space, intersentence space, and so on) consists of "glue" (called *rubber length* in section 9.3.2). Glue has three dimensions: the length of the space, stretchability (the amount by which it can be made longer), and shrinkability (the amount by which it can be made shorter). TEX will stretch and shrink glue and do its best to form lines of equal length.

TEX employs a formula to measure how much stretching and shrinking is necessary in a line. The result is called "badness". Badness 0 is perfect; badness 10,000 is very bad. Lines that are too wide are reported with

```
Overfull \hbox (5.61168pt too wide) in paragraph at lines 49
--57
```

The badness of a line that is stretched too much is reported as follows:

```
Underfull \hbox (badness 1189) in paragraph at lines 93--93
```

Once enough paragraphs are put together, TEX composes a page from the typeset paragraphs using "vertical glue". A short page is marked with a message, for instance:

```
Underfull \vbox (badness 10000) has
occurred while \output is active
```

The typeset file is stored as a dvi (device independent) file.

E.2.3 *Viewing and printing*

Viewing and printing are not really part of TEX, but they are obviously an important part of your work environment. A separate program (called a printer driver) prints the dvi files, and another (the video driver) lets you view them on the monitor.

E.2.4   The files of LaTeX

LaTeX is a "one-pass compiler", that is, it reads the source file only once for type-setting. Therefore, it's necessary for LaTeX to use auxiliary files in which to store information. For the *current* typesetting run, LaTeX uses the auxiliary files compiled during the *last* typesetting run. This explains why you have to typeset *twice* (maybe even three times—see section 6.3.2) to make sure that changes you have made are reflected in the typeset document. These auxiliary files have the same (base) name as the source file; the extension indicates the type of the auxiliary file.

The most important auxiliary file is the aux file. It contains a lot of information, most importantly, the data relevant to symbolic referencing. Here are two typical entries:

```
\newlabel{struct}{{5}{2}}
\bibcite{eM57a}{4}
```

The first entry indicates that a new symbolic reference was introduced in the source file:

```
\label{struct}
```

The command \ref{struct} produces 5, while \pageref{struct} yields 2.

There is an aux file for the source file being processed, and another one for each file included in the main file with an \include command. No aux is written if the \nofiles command is given. The message

```
No auxiliary output files.
```

in the log file reminds you that \nofiles is in effect. The log file contains all the information shown on the monitor during the typesetting. The dvi file contains the typeset version of the source file.

There are five auxiliary files that store information for special tasks. They are written only if that special task is invoked by a command. They are all suppressed if there is a \nofiles command. They are

glo Contains the glossary entries produced by the \glossary commands. A new file is written only if there is a

```
\makeglossary
```

command in the source file (see section 11.5).

idx Contains the index entries produced by the \index commands. A new file is written only if there is a

```
\makeindex
```

command in the source file (see section 11.3).

lof   Contains the entries used to compile a list of figures. A new file is written only if there is a

`\listoffigures`

command in the source file (see section 6.4.3).

lot   Contains the entries used to compile a list of tables. A new file is written only if there is a

`\listoftables`

command in the source file (see section 6.4.3).

toc   Contains the entries used to compile a table of contents. A new file is written only if there is a

`\tableofcontents`

command in the source file (see section 6.3.2).

For the auxiliary files of BIBTEX and *MakeIndex*, see sections 10.2.4 and 11.3. Some classes and packages define other auxiliary files.

PostScript fonts

In section E.1, it was stated that one of the major goals of the new LaTeX was to unify the various dialects of LaTeX. Another major goal was to make it easy to use PostScript fonts.

The Computer Modern fonts were "hard wired" into LaTeX. Many users liked LaTeX but disliked the Computer Modern fonts. With the spread of personal computers and PostScript laser printers, it was imperative that PostScript fonts become integrated into LaTeX. I illustrate with two examples how easy it is to use PostScript fonts with LaTeX 2_ε.

The PSNFSS (PostScript New Font Selection Scheme) distribution, by Sebastian Rahtz, is part of the LaTeX 2_ε distribution. If you do not have it, get it from your TeX supplier or CTAN (see section G.2). It contains all the files—but not the fonts themselves—you need for this appendix.

F.1   The Times font and MathTime

As a first example, step through the process of incorporating the Adobe Times font into a LaTeX document to replace the Computer Modern text fonts, and, optionally, of using the *MathTime* math fonts to replace the Computer Modern math fonts.

Recall from section 2.6.2 that a document class specifies three standard font families:

- the roman (or upright and serifed) document font family;
- the sans serif document font family;
- the typewriter style document font family.

The times package (in the PSNFSS distribution) makes Times the roman font family, Helvetica the sans serif font family, and Courier the typewriter style font family.

Setting up Times   First install the Adobe Times, Helvetica, and Courier Post-Script fonts and their TeX font metric files. Now typeset the `psfonts.ins` file (in the PSNFSS distribution) with the LaTeX format. This will produce a `sty` (style) file for the standard PostScript fonts. The Times style file is called `times.sty`. Copy it into your TeX input directory.

To use the times package, you must have the *font definition* (`fd`) files for the fonts involved. By checking the `times.sty` file, you'll see that you need three fonts: Times, Helvetica, and Courier. In the `times` package these are named `ptm`, `phv`, and `pcr`, respectively; these are the font names in the naming scheme introduced by Karl Berry. In `ptm`, `p` stand for the foundry's name (Adobe), `tm` stands for Times, `hv` for Helvetica, and `cr` for Courier. The corresponding font definition files are named `OT1ptm.fd`, `OT1phv.fd`, and `OT1pcr.fd`, respectively. (`OT1` designates the old TeX font encoding scheme, which is not discussed here.) You can get these from the CTAN distribution sites (see section G.2). Place all these font files in your TeX input directory.

Using Times   In the preamble of your document, type

```
\usepackage{times}
```

after the `\documentclass` line, so that Times will become the roman, Helvetica the sans serif, and Courier the typewriter style document font family.

That's all there is to it.

The times package changes the document font family throughout the document, but you may want to switch to Times only occasionally. For example,

```
{\fontfamily{ptm}\selectfont
this is typeset in the Times font}
```

will typeset the phrase in the Times font:

this is typeset in the Times font

The text preceding and following will not be affected. Similarly,

	0	1	2	3	4	5	6	7	8	9
0										
10							ı		`	´
20	˘	ˇ	¯	˚	¸	ß	æ	œ	ø	Æ
30	Œ	Ø		!	”	#	$	%	&	’
40	(	)	*	+	,	-	.	/	0	1
50	2	3	4	5	6	7	8	9	:	;
60	<	=	>	?	@	A	B	C	D	E
70	F	G	H	I	J	K	L	M	N	O
80	P	Q	R	S	T	U	V	W	X	Y
90	Z	[	\	]	^	˙	‘	a	b	c
100	d	e	f	g	h	i	j	k	l	m
110	n	o	p	q	r	s	t	u	v	w
120	x	y	z	{	}	″	~	¨		

Table F.1: Lower font table for the Times font

	0	1	2	3	4	5	6	7	8	9
120									Ä	Å
130	Ç	É	Ñ	Ö	Ü	á	à	â	ä	ã
140	å	ç	é	è	ê	ë	í	ì	î	ï
150	ñ	ó	ò	ô	ö	õ	ú	ù	û	ü
160	†	°	¢	£	§	•	¶		®	©
170	™							±		
180	¥	µ						ª	º	
190			¿	¡	¬		ƒ			«
200	»	…		À	Ã	Õ			–	—
210	“				÷		ÿ	Ÿ	⁄	¤
220	‹	›	fi	fl	‡	·	‚	„	‰	Â
230	Ê	Á	Ë	È	Í	Î	Ï	Ì	Ó	Ô
240		Ò	Ú	Û	Ù					
250					˛					

Table F.2: Upper font table for the Times font

```
\fontfamily{ptm}\selectfont
this is typeset in the Times font
\normalfont
```

will also typeset the same phrase in Times. (The \normalfont command restores the document font family—see section 2.6.2, so this version is different from the previous one.)

The layout of the Times font is shown in Tables F.1 and F.2. As you can see there are plenty of opportunities to use the \symbol command (see section 2.4.4); for instance, you can define

```
\newcommand{\TM}{{\fontfamily{ptm}\selectfont\symbol{170}}}
```

and then \TM prints ™. To obtain these tables, open the fonttbl.tex file in your ftp directory, add the

```
\usepackage{times}
```

line in the preamble, and typeset it with LaTeX format.

Setting up *MathTime* Looking at a mathematical article typeset with the Times text font, you may find that the Computer Modern math symbols look too thin. Michael Spivak modified the CM math symbols to match the Times font (and other PostScript fonts); he named these modified fonts *MathTime*. You can purchase the *MathTime* fonts from Y&Y, (508) 371-3286, e-mail: sales-help@YandY.com.

Install the *MathTime* PostScript fonts and the TEX font metric files for the *MathTime* fonts. Get the mathtime.ins and mathtime.dtx files (written by Aloysius G. Helminck) from the CTAN. Copy them into your TEX input directory. Typeset mathtime.ins with LaTeX format. This will produce the mathtime.sty file and the necessary fd files.

Using *MathTime* If you want to use Times as the document font family and *MathTime* as the default math font, specify

```
\usepackage{times,mathtime}
```

in the preamble of your document.

F.2   LucidaBright fonts

In this section, I show you how to replace the Computer Modern fonts with the *LucidaBright* fonts (both text and math fonts) in a LaTeX document. You can purchase the LucidaBright fonts from Y&Y.

Get the lucida.ins and lucida.dtx files (by Sebastian Rahtz), which are part of the PSNFSS distribution. Copy them into your TEX input directory. Typeset lucida.ins with LaTeX format. This will produce the lucbr.sty file and a large number of fd files.

Now in the preamble of your document, add the line

```
\usepackage[yy]{lucbr}
```

The yy option is necessary for the LucidaBright package from Y&Y.

APPENDIX

G

Getting it

If your computer is connected to the Internet (directly or through a telephone line), then you can obtain the most up-to-date versions of LaTeX and $\mathcal{AMS}$-LaTeX (and any other distributions you wish to use; for instance, PSNFSS). This appendix explains how to do this.

Writing this section so as to satisfy everybody's needs is about as likely to be successful as writing a section on how to install LaTeX ... So I make some assumptions: you know how to "sign on" to an Internet provider (whose computer I shall call the "local computer"), and "download" from the local computer to your computer. I hope, however, that even if your situation is different from what is assumed, there is enough information here to help you.

You should also keep in touch with TUG, the TeX Users Group. The last section tells you how.

G.1 Getting TeX

There are commercial, shareware, and freeware implementations of TeX.

Commercial implementations

Most commercial TEX implementations are *integrated* (that is, the application contains an editor, viewer—or video driver—and printer driver, as well as TEX). Commercial TEX implementations provide technical support, whereas public domain, freeware, and shareware programs most likely do not; for a novice, this may be an important consideration.

On an IBM compatible, the most popular integrated package is PCTEX for Windows,[1] and on a Macintosh it is TEXTURES.[2]

Some users prefer a nonintegrated setup, since then they can use the editor of their choice and the best video and printer driver. Y&Y tex is one such package.[3]

Public domain, freeware, and shareware implementations

There are also a number of public domain, freeware, and shareware TEX implementations. These can be obtained from a variety of sources (inquire from TUG—see section G.7).

Probably, the most popular DOS TEX implementation is emTEX. It is available from CTAN (see section G.2) and also on a CD-ROM (4allTeX) from the Dutch user group NTG (e-mail: `ntg@nic.surfnet.nl`). The most popular shareware implementation on the Macintosh is OzTEX, available from CTAN. UNIX users get the TEX source code (in C) from CTAN, and compile it. A version of UNIX for IBM compatibles, Linux, will automatically install TEX for you.

Most TEX implementations come with LaTEX (version 2e) and the $\mathcal{AMS}$ enhancements ($\mathcal{AMS}$-LaTEX) as part of the package. If your TEX came from such a source, it may be simplest to keep everything up-to-date from the same source.

If you need support, try the `comp.text.tex` newsgroup on the Internet. You could also try the Frequently Asked Questions (FAQ) documents maintained on CTAN (in the `/tex-archive/help` directory).

G.2   Where to get it?

You can find LaTEX and the $\mathcal{AMS}$ enhancements at one of the Comprehensive TEX Archive Network (CTAN) sites: `ftp.shsu.edu` (U.S.), `ftp.tex.ac.uk` (U.K.), and `ftp.dante.de` (Germany). (There are also a number of mirror sites, ten at the time of this writing; check for the one closest to you.) The primary location for $\mathcal{AMS}$-LaTEX is the $\mathcal{AMS}$ file server `e-math.ams.org`.

At the CTAN sites, the following subdirectories may be of immediate interest to you:

---

[1] Personal TEX Inc., (800) 808-7906, e-mail: `pti@crl.com`
[2] Blue Sky Research, (800) 622-8398, e-mail: `sales@bluesky.com`
[3] Y&Y, (508) 371-3286, e-mail: `sales-help@YandY.com`

The LaTeX distribution

```
/tex-archive/macros/latex
```

is the main LaTeX directory. It has a number of subdirectories, including

- base, the current LaTeX distribution;
- packages, which includes the LaTeX tools, PSNFSS, Babel, and graphics;
- contrib, the user-contributed styles and packages;
- unpacked, which contains the LaTeX distribution unpacked.

AMS-LaTeX

The AMS-LaTeX distribution is in the directory

```
/tex-archive/fonts/ams
```

It contains several subdirectories. To install AMS-LaTeX, copy the contents of the following two directories

```
/tex-archive/fonts/ams/amslatex/inputs
/tex-archive/fonts/ams/amsfonts/latex
```

into your TeX input directory. This is all you really have to do.

The amslatex directory contains the document classes and packages. This directory has three subdirectories: math, inputs, and classes. The files that must go into your TeX input directory (the document class and package files) are all in the inputs directory. The documentation is in the other two directories.

Finally, the amsfonts/latex directory contains the font related packages, and the font definition files.

PostScript fonts

First, you need the PSNFSS distribution. You'll find it in the psnfss subdirectory of the directory packages; the full path of this subdirectory is

```
/tex-archive/macros/latex/packages/psnfss
```

There is also a subdirectory called lw35nfss you may want to explore.

Second, you need the font definition files for all the PostScript fonts you intend to use. To find these files, look in the /tex-archive/fonts/metrics directory. All the Adobe PostScript fonts are in the adobe subdirectory. For instance, to find the Times font definition file, change into the times subdirectory; it has a subdirectory fd, in which you'll find OT1ptm.fd. The full name of this directory is

```
/tex-archive/fonts/metrics/adobe/times/fd
```

cd *path*	change directory to *path*
cd ..	move up one level in the directory structure
ls	list files and subdirectories
mkdir *path*	make directory *path*
rm *name*	remove file *name*
pwd	display current path
rmdir *path*	remove directory to *path*

Table G.1: Some UNIX commands

CTAN help

Robin Fairbairns' companion.ctan document (in /tex-archive/info) lists the names of packages, fonts, etc., and where they can be found.

If you can connect to the Internet with a World Wide Web browser (say, Netscape Navigator), try the CTAN-Web Home Page

http://jasper.ora.com/ctan.html   *http://www.ctan.org/*

It's to an ftp connection what a luxury sedan is to a bicycle.

There are now a number of other WWW sites dedicated to LATEX and related topics; the LATEX Navigator (http://www.loria.fr/tex/english/index.html) has a listing under the title "WWW servers dedicated to (La)TeX".

G.3   Getting ready

Follow a very simple strategy:

1. sign on to the local computer and create a subdirectory to copy the files into;
2. ftp to a CTAN site;
3. locate the directory containing the files you want;
4. download those files to the local computer;
5. transfer the files to your personal computer;
6. unpack them;
7. copy them to the appropriate directories.

Since this involves going through this sequence of steps as many times as there are directories to get, you can improve on this strategy by creating a directory for all the CTAN directories you want to copy. However, let me illustrate the simpler strategy.

You'll need the UNIX commands in Table G.1. Once connected to the CTAN site, use the ftp commands in Table G.2.

cd *path*	change directory to *path*
dir	list files and subdirectories
get *name*	get the file *name*
lcd *path*	change directory in the local computer to *path*
mget *	get all the files in the current subdirectory
prompt	do not prompt which files to get
ascii	set file type to ascii
binary	set file type to binary
quit	disconnect and quit ftp program

Table G.2: Some ftp commands

G.4   *Transferring files*

I'll illustrate the procedure outlined in section G.3, by retrieving the LaTeX distribution. In this session, what the computer displays on the monitor is shown
in this style
whereas user input is shown
in this style
In what follows, "%" is the UNIX prompt and "ftp>" is the ftp prompt. Note: these prompts may be different on your system.

Step 1   Create a subdirectory called transfer on the local computer.

% *pwd*
/home/u1/gratzer

% *mkdir transfer*

% *cd transfer*

% *pwd*
/home/u1/gratzer/transfer

Step 2   Use the method called *anonymous ftp* to log in to the (nearest) CTAN site:

% *ftp ftp.shsu.edu*
Connected to PIP.SHSU.EDU.
220 pip.shsu.edu FTP server (Version 2.1aWU(1)
Fri Aug 20 14:31:05 CDT 1993) ready.

Name (ftp.shsu.edu:gratzer): *anonymous*
331 Guest login ok, send your complete e-mail address as password.

Password: *George_Gratzer@umanitoba.ca*
230 Guest login ok, access restrictions apply.

Step 3   Now move to the directory containing the LaTeX distribution:

```
ftp> cd /tex-archive/macros/latex/base
250 CWD command successful.

ftp> pwd
257 "/tex-archive/macros/latex/base" is current directory.
```

Step 4 Transfer all the files contained in this directory:

```
ftp> prompt
Interactive mode off.

ftp> mget *
200 PORT command successful.
150 Opening ASCII mode data connection for 00readme.txt
(1846 bytes).
...
...

ftp> quit
221 Goodbye.
```

Now all the files have been transferred from the CTAN site to the local computer.

Note the command prompt; without it you would be prompted for every file being transferred.

Step 5 The following session is typical. On the local computer, the current directory should be the one containing the files you wish to transfer.

```
% pwd
/home/u1/gratzer/transfer

% kermit
C-Kermit, 4E(072) 24 Jan 89, SUNOS 4.x
Type ?  for help

C-Kermit> send *
Escape back to your local system and give a RECEIVE command...
```

Now follow the instructions of the communication software on your personal computer to receive files sent by Kermit.

Step 6 At this point, all the files you have just transferred are in a directory on your personal computer. The next step, unpacking the files, is dependent on the individual files themselves. The following general comments may help.

- Look for a file called README or some other variant. This document will point out which file contains the instructions for installation, the user guide, and so on.

- As a rule, the txt files are the ones you should read right away. The tex files are probably user guides, and ins files are the installation files.

    The LaTeX installation is an exception; you must use initex for typesetting the unpack.ins document. All the others are to be typeset with LaTeX format.

Step 7   The installation unpacks a large number of files. Some of these should be copied over to the TeX input directory. These files typically have extensions cls (document class files), sty (packages), clo (class options), fd (font definition files), and bst (BibTeX style files). There are a few others you have to copy with extensions such as ltx, def, and even some with tex. The installation instructions give more detail.

G.5   *More advanced file transfer commands*

The procedure described in section G.4 will provide you all the files you need, but not necessarily in the most convenient or efficient manner. Consult a UNIX book or an Internet book for short cuts (see section I.2). However, the following suggestions may help.

1. When you find yourself in an unfamiliar subdirectory, use the

    get *name* -

    command to read short documentation files. For instance, when in a subdirectory with a README file, issue the command:

    ftp> *get README* -

    This will display the contents of README on your monitor. If the file is long, you may want to view it a screenful at a time, so issue the command

    ftp> *get README* - |*more*

    instead.
2. If you know (part of) the name of a file, but you do not know where to find it on CTAN, issue the command

    ftp> *quote site index latex.ltx*

    to locate the latex.ltx file. This will produce a listing of the directories in which the file can be found. Of course, if the file is packed, this method will not work.
3. Read the files

    README.archive-features
    README.site-commands

    for special commands available at CTAN sites.

4. Many of the directories you may be interested in are available as single tar files, marked by the extension tar. Some directories are also compressed, as signified by the extension Z. In addition, at CTAN sites, you can "compress on-the-fly".

Here is a sample session:

```
ftp> pwd
257 "/tex-archive/macros/latex" is current directory.

ftp> binary
200 Type set to I.

ftp> get base.tar.Z
200 PORT command successful.
150 Opening BINARY mode data connection for /bin/tar.
226 Transfer complete.
local: base.tar.Z remote:  base.tar.Z
826681 bytes received in 2.5e+02 seconds (3.2 Kbytes/s)

ftp> quit

% ls -l base*
-rw------- 1 gratzer 826681 Dec 6 16:20 base.tar.Z

% uncompress base.tar.Z

% ls -l base*
-rw------- 1 gratzer 2242560 Dec 6 16:20 base.tar

% tar -xf base.tar

% ls -l base*
-rw------- 1 gratzer 2242560 Dec 6 16:20 base.tar
... base:
total 2226
-rwx------ 1 gratzer 1846 Jun 12 15:04 00readme.txt
...
```

This is followed by a listing of the new base directory.

The first *ls -l base** command yields the information that the compressed tar file has 826,681 characters; the second, that the uncompressed tar file has 2,242,560 characters; the third gives a complete listing (not shown) of the files in the new directory.

You need only two new commands to uncompress and to create the new directory. Observe that as a result of the compression only 826,681 bytes had to travel from the CTAN site to the local computer (rather than 2,242,560); as a result of combining all the files into one, the transmission happened in *one step*.

G.6   *The sample files*

There are number of sample files discussed in this book as discussed on page 4. You may choose to type these files yourself or get them from the ftp server at the University of Manitoba in the `latex/mil` directory. (As a password, type your e-mail address.) Here is a transcript of how to get them from the University of Manitoba:

```
% ftp ftp.cc.umanitoba.ca
Connected to canopus.cc.umanitoba.ca.
220 canopus FTP server.

Name: anonymous
331 Guest login ok, send your complete e-mail address as password.

Password:
230 Guest login ok, access restrictions apply.

ftp> cd latex/mil
250 CWD command successful.

ftp> pwd
257 "/latex/mil" is current directory.

ftp> prompt
Interactive mode off.

ftp> mget *
200 PORT command successful.
150 Opening ASCII mode data connection for amsart.tpl (4598 bytes).
226 Transfer complete.
local:  amsart.tpl remote:  amsart.tpl
4797 bytes received in 0.048 seconds (97 Kbytes/s)
...
...

ftp> quit
221 Goodbye.
```

Now transfer all the files from the local computer to your computer to the `ftp` directory, and read the `readme` document in the `latex` directory. It'll advise you when I'll have a WWW site to supplement the `ftp` site, the addresses of other sites that carry the sample files, and so on.

G.7   *$\mathcal{AMS}$ and the user groups*

The $\mathcal{AMS}$ provides excellent technical advice. You can reach the $\mathcal{AMS}$ technical staff by e-mail: `tech-support@math.ams.org` or by telephone: (800) 321-4267 or (401) 455-4080.

The TẼX User's Group (TUG) does a tremendous job of maintaining TẼX, developing LẼTẼX, and publishing a quarterly journal (*TUGboat*) and newsletter (TẼX and TUG NEWS). Join TUG if you have an interest in TẼX or LẼTẼX. The address of TUG is P.O. Box 869, Santa Barbara, CA 93101; telephone: (805) 899-4673, e-mail: `TUG@tug.org`.

There are a large number of TẼX user associations that are geographic or linguistic in nature. The three largest national user groups are GUTenberg (French), Dante (German), and UK TUG (U.K.); each is represented on the TUG Board of Directors. All the (national) TẼX user associations are listed in the Resource Directory, a supplement to *TUGboat*. For inquiries about TẼX user associations, consult the Resource Directory, or write (e-mail) TUG.

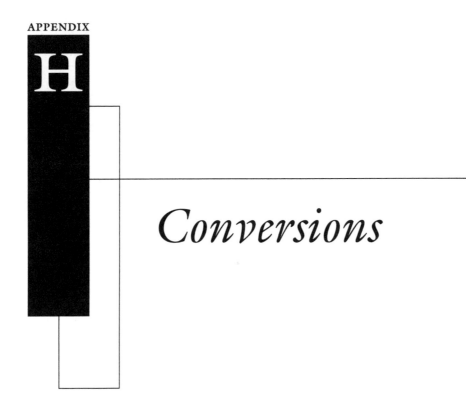

APPENDIX

H

Conversions

There are four groups of experienced users who may want to *convert* to $\mathcal{AMS}$-LaTeX (version 1.2): users of Plain TeX, LaTeX, and $\mathcal{AMS}$-TeX, and users of $\mathcal{AMS}$-LaTeX version 1.1.

Conversion has at least two meanings:

1. Reworking an existing article to utilize LaTeX 2_ε along with the amsmath and amssymb packages (and maybe others).
2. Reworking an existing article to use the amsart document class and the amssymb package (and maybe others).

The two types of conversions are almost identical except that the first type starts with a template file discussed in section 1.6, while the second starts utilizes a template file obtained in section 8.3 by customizing amsart.tpl.

H.1   From Plain TeX

To convert an article from Plain TeX:

- Make sure that you do not use any of the commands listed in section H.1.1.

- Take your personalized article template, save it under a new name, and type in the article information. Then cut the TEX article and paste it into the document environment.
- Replace the TEX displayed math delimiters $$ by \[and \].
- Redo all the section (and subsection) titles as discussed in section 1.7.2, and insert the cross-references.
- Redo the bibliography as shown in section 1.7.4. Insert the cross-references to the bibliographic items.
- Define and invoke all proclamations as described in section 3.4.
- Redesign, if necessary, the multiline formulas using the multiline math environments of the amsmath package.
- Convert tables to tabular form.

H.1.1 TEX code in LATEX

Nearly all document and page design problems are addressed by LATEX or by a package. Therefore, mixing TEX and LATEX is seldom necessary, and requires a deep understanding of the underlying code. Do not mix TEX code in LATEX unless you have mastered the internals of both.

There are a number of reasons why a (Plain) TEX command may not work as expected in LATEX.

- LATEX rewrote the "output" routines of TEX, that is, the way paragraphs are formatted. Avoid all TEX commands that directly affect output (see Chapter 15 of [24]).
- LATEX provides a number of environments that make some TEX commands obsolete: tabbing and center are two examples.
- A number of TEX commands that change the font size are not defined in LATEX.
- Some TEX commands change parameters that are also used by LATEX. For instance, \hangindent in a LATEX list environment will change the shape of the list.

Table H.1 provides a short list of (Plain) TEX commands to avoid.

In addition, the amsmath package prohibits the use of the TEX generalized fraction commands, namely,

```
\over     \overwithdelims
\atop     \atopwithdelims
\above    \abovewithdelims
```

H.2 From LATEX

The conversion depends on whether you start from LATEX version 2.09 or LATEX version 2e.

\+	\fivei	\midinsert	\sevensy
\advancepageno	\fiverm	\nopagenumbers	\tabalign
\beginsection	\fivesy	\normalbottom	\tabsdone
\bye	\folio	\oldstyle	\tabset
\centering	\footline	\pagebody	\tabs
\cleartabs	\footstrut	\pagecontents	\teni
\dosupereject	\headline	\pageinsert	\topins
\endinsert	\leqalignno	\pageno	\topinsert
\end	\line	\plainoutput	\vfootnote
\eqalignno	\magnification	\settabs	
\eqalign	\makefootline	\sevenbf	
\fivebf	\makeheadline	\seveni	

Table H.1: TeX commands to avoid in LaTeX

H.2.1   Version 2e

There are just a few adjustments to make when switching from LaTeX 2_ε.

To do the first type of conversion (see the introduction to this appendix), just add the lines

```
\usepackage{amsmath}
\usepackage{amssymb}
```

after the \documentclass line.

To do the second type of conversion:

- Take your amsart.tpl personalized article template, save it under a new name, and type in the article information. Cut from the LaTeX article the document environment and paste it in the new article.
- Define and invoke all declarations in the form described in section 3.4.
- Do not use a LaTeX bibliographic style if you use an AMS document class; use the AMS version instead.

  With both types of conversion,

- Redesign, as necessary, the multiline formulas using the multiline math environments of the amsmath package. In particular, you should change the LaTeX math environment eqnarray to an amsmath multiline math environment.

H.2.2   Version 2.09

First, you have to convert from LaTeX 2.09 to LaTeX 2_ε. This is usually very easy. In the preamble, you have to change the

```
\documentstyle
```

command to

```
\documentclass
```

Most of the document style options have become packages, which are invoked with the \usepackage command. You may optionally change the two-letter font changing commands (see section 2.6.6) to the corresponding LaTeX 2_ε commands (see section 2.6), but this is not really necessary. Some math symbols that were in LaTeX version 2.09 are now in the package latexsym (see section H.2.3). Now proceed as in section H.2.1.

H.2.3   The LaTeX symbols

The latexsym package defines 11 math symbols. These were originally part of the standard LaTeX setup, but in version 2e they are invoked with the latexsym package. The only frequently used symbol in this set is \Box, which many writers use as the end-of-proof symbol.

If you follow my suggestion and always invoke the amssymb package in your LaTeX document, then you'll have access to all eleven of these symbols, but under different names. Table H.2 presents a rough translation table.

latexsym	amssymb
\mho	\mho
\Join	\bowtie
\Box	\square
\Diamond	\lozenge
\leadsto	\rightsquigarrow
\sqsubset	\sqsubset
\sqsupset	\sqsupset
\lhd	\trianglelefteq
\unlhd	\vartriangleleft
\rhd	\trianglerighteq
\unrhd	\vartriangleright

Table H.2: A translation table

The last four symbols in latexsym are binary operations, while their amssymb equivalents are binary relations. So a more precise translation of, say, \lhd is

```
\mathop{\trianglelefteq}
```

H.3   From A_MS-T_EX

Although the amsmath package is A_MS-T_EX recoded for LaTeX, there are a number of differences the A_MS-T_EX user has to get used to. These seldom cause difficulty since mistakes are caught by LaTeX, as a rule.

The major differences are:

- *AMS*-TeX uses pairs of commands of the form

$$\command \quad \text{and} \quad \endcommand$$

to delimit environments; for instance,

$$\document \quad \text{and} \quad \enddocument,$$
$$\proclaim \quad \text{and} \quad \endproclaim.$$

- Some *AMS*-TeX commands were dropped because there were LaTeX commands that accomplished the same task.
- Some *AMS*-TeX commands became optional parameters of packages or commands.
- Some *AMS*-TeX commands were renamed because there were already LaTeX commands of the same name.
- Bibliographic formatting commands.

So here is what you should do when converting from *AMS*-TeX:

- Take an appropriate personalized article template, save it under a new name, and type in the article information. Then cut the *AMS*-TeX article from \document to \enddocument and paste in the body of the new article.
- Replace the *AMS*-TeX displayed math delimiters $$ by \[and \].
- Look for *AMS*-TeX commands that start with \end. Change all these to the corresponding environment. In particular, redo each proclamation as an environment.
- Completely redo the bibliography. Change the \cite commands to references by labels.
- Redo every user-defined command. Notice that the syntax changes substantially: change \define to \newcommand and \redefine to \renewcommand.

A number of *AMS*-TeX commands that affect the style of the whole document became document class or amsmath package options (and two were dropped). They are listed in Table H.3.

The *AMS*-TeX commands in Table H.4 may cause some difficulties.

H.4   From *AMS*-LaTeX version 1.1

Despite dramatic changes behind the scenes, users of *AMS*-LaTeX version 1.1 have very little to change:

- The \documentstyle command must be changed to \documentclass in the preamble.

A$\mathcal{M}$S-T$_E$X command	A$\mathcal{M}$S-L^{A}T$_E$X
\CenteredTagsOnSplits	centertags document class option
\LimitsOnInts	intlimits amsmath option
\LimitsOnNames	namelimits amsmath option
\LimitsOnSums	sumlimits amsmath option
\NoLimitsOnInts	nointlimits amsmath option
\NoLimitsOnNames	namelimits amsmath option
\NoLimitsOnSums	nosumlimits amsmath option
\TagsAsMath	dropped
\TagsAsText	dropped
\TagsOnLeft	leqno document class option
\TagsOnRight	reqno document class option
\TopOrBottomTagsOnSplits	tbtags document class option

Table H.3: A$\mathcal{M}$S-T$_E$X style commands dropped in A$\mathcal{M}$S-L^{A}T$_E$X

- Most of the document style options became packages, and should be invoked with the \usepackage command. See section 8.4 for the options of the A$\mathcal{M}$S document classes.
- The aligned math environments were redone. Replace the xalignat environment with align, and the xxalignat environment with flalign. The align and flalign environments do not require an argument to specify the number of columns (see section 5.4).
- @ is no longer a special character; typing @ prints @ (make sure that you replace @@ by @ in the \email command of the top matter). As a result, the @- command had to be renamed (it's now \nobreakdash—see section 2.4.8) and the @>>> and @<<< arrow commands can only be used in commutative diagrams (use the \xleftarrow and \xrightarrow commands, see section 4.10.3).
- There are additional commands to change font sizes in version 1.2. You may want to change \small to \Small, \tiny to \Tiny, and \large to \Large (see section 8.1.1).
- The \bold command has been renamed \mathbf (see section 4.14.3).
- The commands

  \newsymbol, \frak, and \BBb

  have been renamed

  \DeclareMathSymbol, \mathfrak, and \mathbb

  They are not provided by the amsmath package but by the amsfonts package (automatically loaded by amssymb).
- The following rarely-used commands have been moved to the amsxtra package:

$\mathcal{AMS}$-TEX	$\mathcal{AMS}$-LATEX
\:	Conflict: renamed \colon
\adjustfootnotemark	Dropped: reset the counter footnote
\and	Renamed \And
\boldkey (math style change)	Dropped: use \boldsymbol
\botsmash	Dropped: use the optional parameter b of \smash
\caption	Changed: use the figure environment and the \caption command
\captionwidth	Dropped: use the figure environment and the \caption command
\cite	Different syntax
\displaybreak	Trap: place it before \\
\dsize (math size change)	Dropped: use \displaystyle
\foldedtext	Dropped: use \parbox
\hdotsfor	Different syntax
\innerhdotsfor	Dropped
\italic (math style change)	Dropped: use \mathit
\midspace	Dropped: use the figure environment
\nopagebreak in multiline math environments	Dropped
\pretend ... \haswidth	Dropped: pad the label with blanks
\roman (math style change)	Conflict: use \mathrm
\slanted (math style change)	Dropped: use \mathsl
\ssize (math size change)	Dropped: use \scriptstyle
\sssize (math size change)	Dropped: use \scriptscriptstyle
\spacehdotsfor	Dropped: use the optional parameter of \hdotsfor
\spaceinnerhdotsfor	Dropped: use the optional parameter of \hdotsfor
\spreadlines	Dropped
\thickfrac	Dropped: use the optional parameter of \frac
\thickfracwithdelims	Dropped: use the optional parameter of \fracwithdelims
\topsmash	Dropped: use the optional parameter of \smash
\topspace	Dropped: use the figure environment
\tsize (math size change)	Dropped: use \textstyle
\vspace in multiline math environments	Dropped: use the optional argument of the \\ command

Table H.4: $\mathcal{AMS}$-TEX commands to avoid

\accentedsymbol,

\sphat, \spcheck, \sptilde, \spdot, \spddot, \spdddot, \spbreve.

See section 4.9 for these commands. If you need any of these commands, add the line

\usepackage{amsxtra}

to the preamble of your document.

- \fracwithdelims has been removed. Code it with \genfrac (see section 4.16).
- \lcfrac and \rcfrac are now options of \cfrac; replace

  \lcfrac with \cfrac[l],   \rcfrac with \cfrac[r]

  (see section 4.16).
- The functionality of the multiline subscript and superscript environments, Sb and Sp, is taken over by the substack command (see section 4.8.2).
- The environments pf and pf* have been renamed proof and proof*, respectively.
- You may wish to redo some of your font commands to the new ones. If you use CM fonts, there is no hurry; the old ones should work as before.

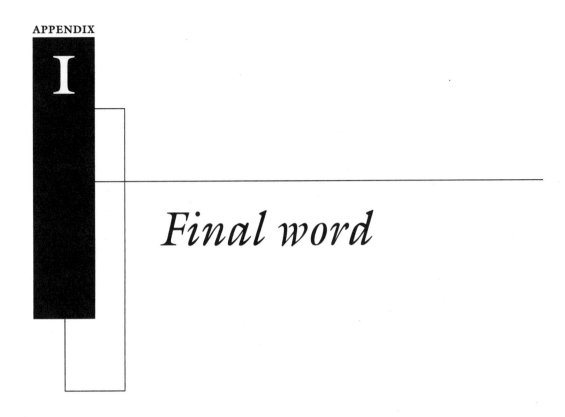

Final word

In this final appendix, I'll outline what was left out of the presentation, and what you might read to learn more about LaTeX.

I.1 *What was left out?*

My goal for this book was to provide the reader with a good foundation in LaTeX and $\mathcal{AMS}$-LaTeX; the visual design of the document is left, to a large degree, to the document class. Keeping in mind this task, what did I omit?

I.1.1 *Omitted from LaTeX*

LaTeX has some features omitted from this book:

1. The `picture` environment is a major feature. It allows you to draw simple pictures with lines and circles.
2. The `array`, `tabular`, and `tabbing` environments have a number of features not discussed in this book.
3. LaTeX makes the style parameters of a document and of most LaTeX constructs available to the user. Very few of them are mentioned in this book.

4. The "low level" commands of NFSS provide finer control over fonts.

If you need these features, this is what you should do:

1. Drawing with the `picture` environment has the advantage of portability. This environment is described in Lamport [30]; many extensions are discussed in Chapter 10 of *The LaTeX Companion*. However, I believe that the correct approach today is to use a drawing program with the capability of saving the graphics in encapsulated PostScript format that can be included in the document using the graphics package (see section 6.4.3).
2. The `tabbing`, `tabular`, and `array` environments—and their extensions—are described in detail in Lamport [30] and Chapter 5 of *The LaTeX Companion*.
3. The style parameters for LaTeX are set in the document class. When the publisher changes the name of the document class, the style parameters are changed to their specifications. If you explicitly change the style parameters in your document, then the journal has no easy way to mark up the source file to conform with their publishing style. If you must change the basic style parameters, advise the editor what was done.
4. There are two types of commands of the New Font Selection Scheme: high-level and lowlevel commands. The latter are, by and large, for style designers and package writers. Nevertheless, those who want to use fonts other than the standard Computer Modern fonts would do well to read Chapters 7 and 11 of *The LaTeX Companion*.

    In this book, I use low level commands of NFSS only in two places: in section 9.5 and Appendix F.

I.1.2   Omitted from TeX

Most all of Plain TeX was omitted from this book. TeX is a powerful programming language; you can design any page layout or formula in TeX. Remember, however, that to change the design features, you should be knowledgeable not only in TeX, but also in document design. Also keep in mind our goal that the journal publisher ought to be able to change the design to its own specifications.

I.2   Further reading

Much documentation is included in the LaTeX and *AMS*-LaTeX distributions; in addition, many third-party packages are well documented. You'll find a lot more documentation on CTAN.

As you have no doubt noticed, there are many references to *The LaTeX Companion* in this book. For instance, a whole chapter (Chapter 4) in the *Companion* deals with the layout of the page, and several sections (in Chapter 2) show how to change section headers.

The LaTeX Companion is not a beginner's book, but it's indispensable for advanced LaTeX users with special needs not served by document classes. It is also the best overview of more than a hundred important packages. For package writers and students of NFSS, the *Companion* is *the* basic textbook.

It is a bit more complicated to learn TeX. You may want to start out with Wynter Snow [42]. It introduces many of the basic concepts of TeX in a very relaxed style with many examples; the LaTeX notes make the book especially useful, and the author gives many examples of macros. TeX as a programming language is not discussed, however.

Raymond Seroul and Silvio Levy [41] give another good introduction; this book also has a chapter on TeX programming. Donald E. Knuth [24] provides an easy introduction to TeX, as long as you avoid the difficult parts marked by "dangerous bend signs". Paul W. Abrahams, Karl Berry, and Kathryn A. Hargreaves [1] explain many TeX commands grouped by topic. It has a very useful, nonsequential approach. Victor Eijkhout [11] is an excellent reference book on TeX mainly for experts. For many tutorial examples, see the articles and columns in *TUGboat*.

For advice to authors of mathematical articles, see Ellen Swanson [44]; it is interesting to note how many of Swanson's rules have been incorporated into LaTeX. The points of view on copy editing of the Cambridge and Oxford University Presses are presented in Judith Butcher [8] and Horace Hart [22], respectively.

Ruari McLean [31] gives a useful introduction to typography, the art of printing with type. See also Alison Black [7] for more about typefaces.

Harley Hahn [21] provides an excellent introduction to Unix; and Ed Krol [29] is a good first book on the Internet.

LaTeX focuses on the system-independent nature of TeX. However, everyone works with a *specific* computer system, and many tools are available that make one's labor easier. Norman Walsh [45] is an excellent overview of system-specific tools.

Bibliography

[1] Paul W. Abrahams, Karl Berry, and Kathryn A. Hargreaves, *TEX for the Impatient*. Addison-Wesley, Reading, Massachusetts, 1990.

[2] American Mathematical Society, *AMSFonts version 2.1—Installation Guide*. Providence, RI, 1991.

[3] ——, *AMSFonts version 2.1—User's Guide*, Providence, RI, 1991.

[4] ——, *AMS-LATEX version 1.1—User's Guide*. Providence, RI, 1991.

[5] ——, *AMS-LATEX version 1.2—User's Guide*. Providence, RI, 1995. (For additional information, typeset `amsmath.dtx` and the other `dtx` documents in the AMS distribution.)

[6] ——, *Installation Guide for AMS-LATEX version 1.1*. Providence, RI, 1991.

[7] Alison Black, *Typefaces for desktop publishing: a user guide*. Architecture Design and Technology Press, London, 1990.

[8] Judith Butcher, *Copy editing: the Cambridge handbook*. 2nd ed., Cambridge University Press, London, 1981.

[9] Pehong Chen and Michael A. Harrison, *Index preparation and processing*. Software-Practice and Experience **19** (9) (1988), 897–915

[10] Michael Doob, *TEX Starting from* $\boxed{1}$. Springer Verlag, Berlin, New York, 1993.

[11] Victor Eijkhout, *TEX by topic: A TEXnician's reference*. Addison-Wesley, Reading, Massachusetts, 1991.

[12] Michel Goossens, Frank Mittelbach, and Alexander Samarin, *The LATEX Companion*. Addison-Wesley, Reading, Massachusetts, 1994.

[13] George Grätzer, *Math into TEX: A simple introduction to AMS-LATEX*. Birkhäuser Boston, Boston, 1993.

[14] ——, *AMS-LATEX*. Notices Amer. Math. Soc. **40** (1993), 148–150.

[15] _____, *Advances in TEX implementations. I. PostScript fonts.* Notices Amer. Math. Soc. **40** (1993), 834–838.

[16] _____, *Advances in TEX implementations. II. Integrated environments.* Notices Amer. Math. Soc. **41** (1994), 106–111.

[17] _____, *Advances in TEX implementations. III. A new version of LATEX, finally.* Notices Amer. Math. Soc. **41** (1994), 611–615.

[18] _____, *Advances in TEX. IV. Header and footer control in LATEX.* Notices Amer. Math. Soc. **41** (1994), 772–777.

[19] _____, *Advances in TEX. V. Using text fonts in the new standard LATEX.* Notices Amer. Math. Soc. **41** (1994), 927–929.

[20] _____, *Advances in TEX. VI. Using math fonts in the new standard LATEX.* Notices Amer. Math. Soc. **41** (1994), 1164–1165.

[21] Harle Hahn, *A Student's Guide to Unix.* McGraw-Hill, New York, 1993.

[22] Horace Hart, *Hart's Rules For Compositors and Readers at the University Press, Oxford.* Oxford University Press, Oxford, 1991.

[23] Donald E. Knuth, *The Art of Computer Programming.* Volumes 1–. Addison-Wesley, Reading, Massachusetts, 1968–.

[24] _____, *The TEXbook.* Computers and Typesetting. Vol. A. Addison-Wesley, Reading, Massachusetts, 1984, 1990.

[25] _____, *TEX: The Program.* Computers and Typesetting. Vol. B. Addison-Wesley, Reading, Massachusetts, 1986.

[26] _____, *The Metafontbook.* Computers and Typesetting. Vol. C. Addison-Wesley, Reading, Massachusetts, 1986.

[27] _____, *METAFONT: The Program.* Computers and Typesetting. Vol. D. Addison-Wesley, Reading, Massachusetts, 1986.

[28] _____, *Computer Modern Typefaces.* Computers and Typesetting. Vol. E. Addison-Wesley, Reading, Massachusetts, 1987.

[29] Ed Krol, *The Whole Internet: User Guide & Catalog.* O'Reilly & Associates, Sebastopol, CA, 1992.

[30] Leslie Lamport, *LATEX: A Document Preparation System.* Addison-Wesley, Reading, Massachusetts. First ed. 1985. Second ed. 1994.

[31] Ruari McLean, *The Thames and Hudson Manual of Typography.* Thames and Hudson, London, 1980.

[32] Frank Mittelbach, *An extension of the LATEX theorem environment.* TUGboat **10** (1989), 416-426.

[33] Frank Mittelbach and Chris Rowley, *LATEX 2.09 → LATEX3.* TUGboat, **13** (1) (1992), 96–101.

[34] _____, *LATEX 2ε—A New Version of LATEX*. TEX and TUG NEWS, **2** (4) (1993), 10–11.

[35] _____, *The LATEX3 Project*. Euromath Bulletin, **1** (1994), 117–125.

[36] _____, *LATEX3 in '93*. TEX and TUG NEWS, **3** (1) (1994), 7–11.

[37] Frank Mittelbach and Rainer Schöpf, *The new font family selection—user interface to standard LATEX*. TUGboat **11** (1990), 297–305.

[38] Oren Patashnik, *BIBTEXing*. Document in the BIBTEX distribution.

[39] _____, *BIBTEX 1.0*. TUGboat **15** (1994), 269–273.

[40] Rainer Schöpf, *A new implementation of the LATEX* verbatim *and* verbatim* *environments*. TUGboat **11** (1990), 284-296.

[41] Raymond Seroul and Silvio Levy, *A beginner's book of TEX*. Springer-Verlag, New York, 1991.

[42] Wynter Snow, *TEX for the beginner*. Addison-Wesley, Reading, Massachusetts, 1992.

[43] Michael Spivak, *The Joy of TEX*. 2nd ed. American Mathematical Society, Providence, R.I., 1990.

[44] Ellen Swanson, *Mathematics into Type: Copy Editing and Proofreading of Mathematics for Editorial Assistants and Authors*. American Mathematical Society, Providence, RI, 1986.

[45] Norman Walsh, *Making TEX work*. O'Reilly & Associates, Sebastopol, CA, 1994.

Afterword

This book is based on my earlier book *Math into T_EX: A simple introduction to* $\mathcal{AMS}$-*L*^A*T*_E*X* [13]. Although the topic changed considerably, I borrowed a fair amount of material from that book.

So it may be appropriate to begin by thanking here those who helped me with the earlier book. Harry Lakser was extremely generous with his time; Michael Doob and Craig Platt assisted me with T_EX and UNIX; and David Kelly and Arthur Gerhard read and commented on an early version of that manuscript. Michael Downes, Frank Mittelbach, and Ralph Freese read various drafts. Richard Ribstein read the third and fourth drafts very conscientiously.

The first draft of this new book was read for the publisher by

- David Carlisle (of the L^AT_EX3 team)
- Michael J. Downes (the project leader of the $\mathcal{AMS}$ team)
- Fernando Q. Gouvêa (Colby College)
- Frank Mittelbach (the project leader of the L^AT_EX3 team)
- Tobias Oetiker (De Montfort University)
- Nico A. F. M. Poppelier (Elsevier Science Publishers)

Together they produced a huge tutorial on L^AT_EX for my benefit. I hope that I succeeded in passing on to you some of what I learned from them.

On February 8, 1995, a short announcement was posted on the Internet (in the `comp.text.tex` newsgroup) asking for volunteers to read the first draft. The response was overwhelming.

I received reports from the following volunteers:

- Jeff Adler (University of Chicago, Chicago, IL, USA)
- Helmer Aslaksen (National University of Singapore, Singapore, Republic of Singapore)
- Andrew Caird (University of Michigan Center for Parallel Computing, Ann Arbor, MI, USA)

- Michael Carley (Trinity College, Dublin, Ireland)
- Miroslav Dont (Czech Technical University, Prague, The Czech Republic)
- Simon P. Eveson (University of York, Heslington, York, England)
- Weiqi Gao (St. Louis, MO, USA)
- Suleyman Guleyupoglu (Concurrent Technologies Corporation, Johnstown, PA, USA)
- Peter Gruter (Laboratoir Kastler Brossel, Paris, France)
- Chris F.W. Hendriks (National Aerospace Laboratory, Amsterdam, The Netherlands)
- Mark Higgins (Global Seismology, British Geological Survey, Edinburgh, Scotland)
- Zhihui Huang (University of Michigan, Ann Arbor, MI, USA)
- David M. Jones (Information and Computation, MIT Laboratory for Computer Science, Cambridge, MA, USA)
- Alexis Kotte and John van der Koijk (University Hospital Utrecht, Utrecht, The Netherlands)
- Donal Lyons (Trinity College, Dublin, Ireland)
- Michael Lykke (Roskild, Denmark)
- Steve Niu (University of Toronto, Toronto, ON, Canada)
- Piet van Oostrum (Utrecht University, Utrecht, The Netherlands)
- Denis Roegel (CRIN, Nancy, France)
- Kevin Ruland (Washington University, St. Louis, MO, USA)
- Thomas R. Scavo (Syracuse, NY, USA)
- Peter Schmitt (University of Vienna, Vienna, Austria)
- Nandor Sieben (Arizona State University, Tempe, AZ, USA)
- Paul Thompson (Case Western Reserve University, Cleveland, OH, USA)
- Ronald M. Tol (University of Groningen, Groningen, The Netherlands)
- Ernst U. Wallenborn (Federal Institute of Technology, Zurich, Switzerland)
- Doug Webb (Knoxville, TN, USA)

There were many volunteers, ranging in expertise from users who wanted to learn more about LaTeX, to experts in charge of large LaTeX installations, to internationally known experts whose names are known to many LaTeX users; and ranging in background from graduate students, to professional mathematicians, computer scientists, chemical engineers, psychiatrists, and consultants. I would like to thank them all for their enthusiastic reports. They shared their learning and their teaching experiences. This has become a much better book for their contributions.

I also received carefully crafted reports on Chapter 10 from Oren Patashnik (Stanford University), the author of BibTeX.

Based on these reports (ranging in size from two pages to over thirty pages), the manuscript has been rewritten, most of it has been reorganized, and sections have been added or deleted. I felt that as a result of the major changes probably many new errors have been introduced. So the second draft was again sent to three readers; Jeff Adler, Simon P. Eveson, and David M. Jones sent me 27 pages of reports, confirming my suspicions. My deepest appreciation to these three individuals for their excellent repeat performance.

In the meanwhile, Merry Obrecht Sawdey undertook the visual design of the book. She also lent a helping hand in the final typesetting of the book.

The manuscript then was sent to the technical editor, Thomas R. Scavo, who flooded it with red ink; it would be hard to overstate the importance of his work. The final version of the book was checked again by him.

Last but not least, I want to thank Edwin Beschler, who believed in the project from the very beginning.

Index